Designing for Care

Edited by Jerod Quinn, Martha Fay Burtis, and Surita Jhangiani
Foreword by Catherine J. Denial

Hybrid Pedagogy Inc.
2022

© 2022 Hybrid Pedagogy
Chapters © their respective authors

Published by Hybrid Pedagogy Inc.
https://hybridpedagogy.org
Denver, CO

Cover art by Pawel Czerwinski
Book design by Martha Fay Burtis

Designing for Care by Hybrid Pedagogy Inc. is licensed under a Creative Commons Attribution-NonCommercial 4.0 International License, except where otherwise noted.

First printing, August 2022
ISBN 979-8-9866764-0-1

Contents

Acknowledgements	v
Foreword Catherine J. Denial	vii
Introduction Jerod Quinn	xi
Click, Click, Connect Jessica O'Reilly	1
Intentionally Equitable Hospitality as Critical Instructional Design Maha Bali and Mia Zamora	11
Humanizing Online Learning Mandi Singleton and Nicolas Pares	33
Developing Critical Student Autonomy in Blended Learning Laurel Staab and Martin Wairimu	49
Designing for Inclusion Jennifer Hardwick, Fiona Whittington-Walsh, Kya Bezanson, Anju Miller, and Emma Sawatzky	67
The Straight and Narrow is the Path of Least Resistance Pat Lockley	85
Access Alone Isn't Enough Benjamin D. Remillard	95
Sharing Instructional Design Mary Klann, Logan Gorkov, and Rossel-Joyce Garcia	113
Feeling (Un)Seen Andrew David King	135
Author Biographies	155

Acknowledgements

The editors of this collection would like to thank, first and foremost, the authors who so graciously shared their work with us for this collection. Thank you for staying with us through the call for chapters, months of editing and revisions, a change in editorial lineup, and the long and drawn out process that is book creation. We are deeply proud of this collection, and we hope you are too.

We would also like to specifically thank Sean Michael Morris for igniting the community around critical instructional design, and for seeing a space for this conversation in putting out the call for chapters through Hybrid Pedagogy. Thank you also for your help selecting chapters, for those first rounds of edits, and for handing over this project to us and trusting us to see it through to the end.

Thank you to Jesse Stommel as well, for your production guidance and all the wrangling of Pressbooks and the publication process. We appreciate the help and editing advice you have offered, along with your time and energy on this project. Also, the Zoom cameos from Hazel are always a welcome addition to any meeting.

Thank you, Sukaina Walji, for your help in selecting the chapters for this collection along with the early rounds of edits and revisions. Your assistance and expertise were greatly appreciated.

Jerod would like to also thank the following amazing people for their help, willingness to listen, high-fives, and general encouragement: Amy Archambault, Amanda Stafford, Fatemeh Mardi, Tammy M. McCoy, Chris Grabau, Allen Brown, Donohon Abdugafurova, Norah Elmagraby, and Matt Easter. Thank you to my family as well, especially Charity, who spent many nights helping me think through the dynamics and details of book editing along with helping me to find constructive yet kind ways to say, "I need you to cut 1000 words from this chapter because it's too freakin' long." And while my kids didn't have much to do with this collection, they like seeing their names in books so here you go Grace and Ash.

Surita would like to thank Jerod and Martha for being such amazing editors to work alongside with. I am grateful for the opportunity to have been a part of this project. I learned so much along the way about unique approaches and issues related to critical design and care.

Martha would like to thank Robin, Hannah, and Matt for helping her survive the last three years; Susan and John Fay for being the first people to teach her why writing (and editing!) matters; and Erik, Madigan, and Graeme for being the reasons everything else matters.

Foreword

Catherine J. Denial

I write this at my dining-room table, the locus of my office for most of the past two years. It's year three of the global COVID-19 pandemic, and our lives are divided into six-foot blocks of attention, measured by social distance (and its lack). We have traded cloth masks for N95s (or perhaps no mask at all); we have been vaccinated and boosted (or we have not). We juggle the personal calculus of risk, every decision to travel, to eat out, or to stay home colored by the hope of good health in the face of adversity. Collectively we have not agreed to protect the most vulnerable among us, nor have our governments stepped into that gap. We struggle with financial uncertainty, the vagaries of technology, the limits of broadband, and the scope of what we can claim to know. We are tired and uncertain, we are often overworked and feel underappreciated, and many of us are angry. Our students are too.

We have lost a great deal to the pandemic, including the lives of people we love and, in many cases, our own good health. Many of us lost our footing in teaching when we went online for the first time, or as seasoned online instructors, when we worked with students who had not freely chosen that modality as the one in which they preferred to learn. We craved connection and struggled to know how to create it; we wrestled with new software and hardware to reach students without tipping ourselves or them into overwhelm. All of this has happened at speed. We have largely not had the breathing room to process these realities, absorbed as we have been in the work of adapting to changing circumstances, institutional policies, and personal needs. Too often we have been told that things are returning to "normal," but in truth there is no return to how things were in 2019 (even if that were a desirable destination).

But that is not the end of the story.

By prioritizing care and community in our teaching and instructional design, we give ourselves and our students the opportunity to remake this uncertain world. Our students are thirsty for this, and we teachers no less. All education is relational; all of us deserve support, concern, and kindness, and the chance to work—as these essays demonstrate—in collaboration with each other so that learning can take place. We must ditch what Paulo Freire called the banking model of education. Instead of acting as sages-on-the-stage, depositing wisdom into the passive minds of students, we must (as the authors of these essays do) envision (and actualize) education as a multi-faceted collaboration between students, teachers, and designers. No learning relationship in this book is unidirectional—feedback flows from person to person as everyone involved in the enterprise of education refines their understanding of the work at hand.

These essays offer hope. Hope, as Mariame Kaba famously puts it, is a discipline. It is not simply a feeling but a practice: we must do hope, not just wait for it to materialize. Even in the face of the significant challenges of these pandemic years, we can make hope happen through our actions, and this volume is chock full of suggestions for exactly how. Here you'll find critical reflections on things that have worked in online and hybrid classrooms, and things that have not. You'll find suggestions for making concrete changes—creating a genuine welcome for students in your classroom space; rethinking the design arc of a course; experimenting with digital tools; and transforming assessment. You'll find critiques of our educational systems, and powerful invitations to imagine better. These essays offer each of us valuable companionship in the work of mentoring, design work, teaching, and learning.

Each of these essays offers a vision of education rooted in the building of communities. Sometimes these are communities of two, as an instructor and student sit down at either end of a Zoom connection and share their vulnerabilities. Sometimes these are communities of four, as students work, peer-to-peer, to support one another even as they work on individual projects. Sometimes these communities encompass teachers and students in an entire class, thinking seriously (and joyfully!) about what it means to disrupt the power dynamics of higher ed. In every essay in the following pages you'll find evidence of just how vital community is, and what a powerful difference it makes to those involved.

My own experience as an instructor bears out the wisdom of these essays. It was the online communities of which I am a part, especially on Twitter, that made it possible for me to teach in a pandemic and more, to discover new ways to flourish. In March and April of 2020, I learned from online-education experts about (what seemed at the time to be the mysteries of) asynchronous learning, re-imagining the use of my college's learning management system, and using Zoom. I threw questions out into the Twitter-verse, and people (some of whom I already knew, many of whom I didn't) provided answers. As the pandemic rolled on, I found understanding online, and solidarity in my struggles, as well as a community of people ready to cheer for the things that went right. I experimented with new course designs, formative assessments, and ungrading at large, and I asked my students what they wanted from their classes, and I implemented change. Along the way I reported back to my communities how things were going, began to offer answers to questions myself, and saw the connective tissue of human relationship strengthened by bits and bytes.

If you're looking for inspiration, you will find it here. If you're seeking analysis and data, these essays offer both. If you want new ideas for classroom activities, projects, and assessments, these authors have many to share. But most of all I hope you find in these pages a wealth of connections to the community of which you are a part as an educator and/or learner, and the opportunity to be seen, to be encouraged, and to be understood.

Introduction

Jerod Quinn

Imagining better pedagogies is the first step in creating powerful learning environments. As you will see in the pages of this collection, it is far from the last step. It's certainly a good place to start, though. In March of 2021, *Hybrid Pedagogy* put out a call for chapters for a Critical Instructional Design Reader. That call asked the questions:

> "What if technology had misled us, distracted us from what's actually important for teaching online? What if technology has so far interpreted instruction for us—even from the days of correspondence courses—making the page, digital or otherwise, a surrogate for our pedagogies? How do we reclaim the relational, communal, intimate side of teaching when glass and pixels and apps stand between? When we undertake the work of defining and investigating critical instructional design, we must shift our focus from the screen to the student, from best practices to humanizing pedagogies."

Submissions from North America, Europe, Africa, and Australia came in with a wide scope for how and why a problem-posing approach of critical pedagogy can be applied to online classes. We heard from instructional designers, educators, and students themselves. It also quickly became clear that while the COVID-19 global pandemic did not instigate these conversations, it certainly poured gasoline on the fire of implementing them. Care in online classes began to become a mainstream conversation among all kinds of educators as the pandemic created new tensions and exacerbated old ones at a literal global scale.

As the editors were sifting through the submissions we noticed two related, but divergent streams emerging. One stream was focused on creating environments and experiences grounded in care and compassion. It includes conversations of the logistical hurdles of building intentional hospitality into online experiences, but also the rewards of learners being empowered in the experience. The other stream focused on applying the ideals of critical instructional design to the course design process. These chapters challenged the assumptions of linear, western approaches to higher education and pushed the boundaries of what online learning can look like. The editors made the decision to gather these chapters into two sibling collections: *Designing for Care* and

Towards a Critical Instructional Design. The collection you are reading is the *Designing for Care* edited collection.

Designing for Care

When we talk about designing for care, we are talking about creating, crafting, and teaching our online courses (and all of our courses increasingly mediated by technology) in the intentional narrative of a shared humanity. It's not feminine, masculine, or even gendered work. It's the work of treating people like the fantastic, curious, unpredictable, capable, and multi-layered people they are. It's refusing to see people as one dimensional, regardless of any system-assigned labels like student, teacher, instructional designer, disabled, at-risk, first-generation, or whatever. It's not just being compassionate towards our learners. Being kind to our learners is great, but designing for care is the intentional framing of course design and teaching through a structure that demonstrates care towards all those involved.

There is certainly a resistance to care in education. It's easy to dismiss care as a factor in the classroom to the realm of small, liberal arts classes in small, liberal arts colleges. It's also lazy thinking to do so. Perhaps the opposite of designing for care is designing for efficiency? Not caring is an amazingly efficient practice. There's no lying awake at night thinking about learners, there's no worries about the zoom screen that is not nor will ever be "camera on," there's no need to critique or, heaven forbid, change course design practices. Not caring is amazingly efficient. It's also not an option for educators who take a "problem posing" approach to humanizing online learning.

In the following chapters of this collection you will read how many people across many institutions have worked to design their online learning to be more humanizing to both the learners and the instructors. Some of these designs are theoretical, and some are grounded in practical application. Some worked splendidly, and some had aspects that fell painfully flat. You will hear from people who have been teaching and learning online for decades, as well as people whose first online learning experience came from being forced into a remote teaching context as COVID began raging globally.

This is a wonderfully wide collection of voices from various parts of the world. As editors, we made it our work to create a platform for the authors to share their stories in their voices. One thing you might notice about how we chose to share stories is the vocabulary and spell-

Introduction

ing of certain words. We have authors from Australia, Africa, Europe, and North America all sharing their work. We decided that changing the spelling of words to be the Americanized versions seemed to work against the value of letting people share their own stories in their own words. It might be a "little" thing, but it didn't feel right to change these so you will see spelling variants throughout the collection. That choice was intentional.

Flow of the Collection

The collection is not sorted into themes but there is a narrative flow to the chapters. They are arranged to weave back and forth between the more personal and the academic-leaning writing to create a reading experience that is constantly moving you forward through the twists and turns of the story. No matter the flow, you can still absolutely read them as individual pieces in any order you find helpful.

Click, Click, Connect begins our collection by reminding the reader of the real and felt consequences, both of the learners and the instructors, of a course design grounded in care. These short vignettes tell the stories of our design choices. Not every learner will notice or appreciate this approach, but for many learners designing for care leads to empowerment and a better future.

Intentionally Equitable Hospitality centers values rather than focusing on predetermined, measurable outcomes, and how those values are not neutral. This chapter gives direction on how to create brave spaces for blended connections while decentralizing power in online learning.

Humanizing Online Learning begins by posing the question of, "What does the experience of humanized online learning feel like for instructors and students?" From there they use stories and practical approaches to give examples of designing for care and connection that push the reader to change their design so that it flows from compassion first.

Developing Critical Student Autonomy focuses on partnerships between instructors and learners and how learner autonomy can shape the experience of a course. This chapter is co-written by the instructor and a learner from the course under examination, which provides an amazing window into the perspective of both groups involved. It also shows how to both talk the talk, and walk the walk of instructor-learner partnerships.

Designing for Inclusion is a case study on Including All Citizens in not just an individual course, but in a university program that gives learners with intellectual and/or developmental disabilities the right to an inclusive, flexible, and transformational college education. You will hear from both the instructors guiding the program, and from some of the learners who have lived the program as they share their stories and offer guidance on implementation.

The Straight and Narrow is a bit of an exercise in wordplay, but wordplay that cuts to the bone in a bouncing, three column arrangement.

Access Alone isn't Enough brings the reader face-to-face with the realities of the digital divide, that our learners do not always have access to high-speed connections or the latest and greatest gadgetry. But that doesn't confine us to the realm of Microsoft Documents and plain-text for helping our learners develop more modern digital skills. We can use a careful and caring design to help close some of those opportunity gaps that plague teaching and learning technology.

Sharing Instructional Design is a co-written piece where the process of sharing instructional design between learners and instructors is detailed (and annotated) around a Digital History and Memory course. Sharing design also requires sharing power, and the authors do not shy away from addressing the tensions that can create for everyone involved.

Feeling (Un)Seen is the final chapter in the collection and it gives practitioners a window into the experience of instructors and learners with both hidden and "legible" disabilities. The author challenges educators that designing for the maximally able-bodied person forces learners with disabilities to engage in the often painful work of outing themselves to a culture that tends to exclude and erase them. Designing for care is an approach that creates experiences that provide the flexibility for learners to make themselves only as legible as they choose to be to the university.

Thank you for listening to the voices of the people and the communities who are here to share their work in designing for care with fellow critical instructional designers and educators. We hope this work is both inspirational and challenging to your course design and teaching approaches. If you enjoy this collection, please consider checking out the sibling collection, ***Towards a Critical Instructional Design***.

Click, Click, Connect

Jessica O'Reilly

Preamble

Life is story, and stories hold truths. We can learn so much if we're able to listen.

The following vignettes are based on events that occurred during the 2020-2021 academic year, a pandemic year that called me and so many others to "pivot" to teaching and learning completely online. I hope that these stories serve as a reminder of the many ways that our human connections move through our screens and through us.

– Jess

"I'm Proud of You"

I sip my coffee, waiting for a student to join our virtual meeting. Ding! Ding! I click to admit. One of my cats jumps in my lap as she's joining; his tail wraps around my neck. A fluffy orange scarf. "Morning!" I say, rearranging my little helper so he's comfortable in my lap. "Hey Jess." She looks tired. Her voice is thin, lacking energy. Her skin is pale, hair dishevelled. I scan the room behind her. Unmade bed. Dirty dishes on the dresser. Drawers open, a few clothes peaking out. I take a sip of coffee then ask, "What's up?"

"Who's this?" she asks, referring to the cat. We talked about pets for a while. I meet her puppy. I go get my other cat and together we marvel at his obesity. We swap fur-baby stories. At one point, I put my hand close to my camera to show her the scar my pet hamster gave me decades ago. We joke about being "foster fails." We have an easy rapport, but there's a heaviness too. She's laughing, but it's forced, thin.

Eventually, she says, "I wanted to talk about my paper."

But she doesn't. Instead, she tells me about her life. About how she moved out at sixteen because her parents were both on destructive paths. She shares details. I nod and sip my coffee. She tells me that she's the first in her family to go to college, that she's with a good person who loves her. Whose family loves her. She says it's the first time she's been a part of a healthy family dynamic. The first time she's witnessed encouraging parents. My heart sinks. I keep listening. And caffeinating. I tell my eyebrows not to make those furrow lines that my mum calls "my elevens." I watch myself in the Zoom window. Stay neutral. Use validating phrases. Stop taking so many sips of coffee. Don't be drawing comparisons with your own life. Listen. Listen.

I ignore my inner voice and tell her about my mum. How, like her, my mum sought to break the cycle. How, at seventeen, she got kicked out of the house for getting pregnant with me. How she moved hundreds of kilometres away to finish out her pregnancy at Bethesda Centre for Unwed Mothers. "Justin Bieber's mum went there too," I add. She offers a fragile smile and a slight nod. I explain that my mum finished high school through correspondence courses and evening classes, carting me in my stroller and her acoustic guitar with her on the city bus. I share how eventually, my mum went on to complete two college diplomas and found meaningful work in the developmental sector, helping others. "I see my mum's story in yours," I said. "Do you?"

She nods again, with a bit more vigour this time and shares more about her life, about her dreams and ambitions, her path. She pauses and says quietly, "The trouble is... drugs keep killing people I care about. My cousin died of an overdose last night. We were close. He's my friend. Well, was..." she takes a deep breath. "I think I failed my math test."

I'm surprised she's talked about her math test; I'm her English teacher. My mind runs a quick race. It's midterm, so that was probably an important test. Drugs keep killing her loved ones? I think of the many crosses lined up in my city's downtown core. A memorial to those lost to addiction. The crosses seem to multiply daily. I imagine her standing in front of one, alone.

I offer my condolences. It sounds canned, trite. I choose honesty. "I'm not sure what to say, other than I'm so, so sorry." She continues to share, telling me that an hour after she found out she'd lost her cousin to an opioid overdose, she attempted to write the midterm exam in her math course. She says she doesn't want any special treatment. She wants to be like any other student, but she fears she didn't do well on the test. "Now what should I do?" she asks.

I say what I hope are the right things. Encourage her to self-advocate with the right people. Ask how I can help. We keep talking. I find myself sharing more of my story with her. Time passes. As we're wrapping up, I pause, then awkwardly say, "I'm not sure how to say this... It might be kind of dorky, and I know we haven't known each other that long... but... I just want to tell you that I'm really proud of you. For taking this path, for working through all this stuff. I can't imagine how difficult it's been, but you're making a life for yourself, you're making it happen. You're grieving, but through your grief, you're still trying to get it all done. That's really amazing. Really... brave."

I wish I had better words. I want to acknowledge how profoundly inspiring her commitment to her future self is.

She puts her hood up over her head, way up so that it's almost covering her eyes and says, "Oh I'm not good with the feels.... But, I don't think it's dorky. I never heard 'I'm proud of you' growing up." We sit together for a few more moments, simply being in each other's company.

We never do talk about her paper.

"That's Not Your Job"

My employer is offering a virtual training session. It's the first day of a two-day online course. The facilitator has just finished a mini-lecture on active listening strategies. Now he's unceremoniously launched us into breakout rooms "to discuss." I find it jarring to be blasted through virtual space and plunked into a random "room" with random people. I wait for others to join, smile and exhale when I see a friend is in my group. Then I feel a sinking sensation when I see a certain person pop into the space. Interruptor.

Interruptor is a manager whom I strongly dislike. I am biased toward this person. I can FEEL my bias towards this person in the way my hands and jaw clench when his name appears on my screen. I've had negative interactions with him in the past. While I'm grateful that we aren't grouped together in person, I also feel my interest in the training fading away. What if I just quit and blame a bad internet connection...I hover my mouse over the "leave meeting" button, but think better of it.

I stick around despite my discomfort. As the group gets situated, I remember how earlier in the day, Interruptor held to his nickname, several times. He only interrupted females though. I kept track. I was mid-sentence when he unmuted, spoke over me, dismissed what I was saying, then moved the conversation along to something he deemed more important. The facilitator did not correct him.

I remembered how the blood moved up into my face. My heart did an angry BOOM BOOM beat as I glared at his little Zoom square. I hit my mute button in frustration, then texted my friend who was also on the call. I texted angry stuff that shouldn't be repeated. She joked and calmed me down. But I still felt embarrassed. Silenced. And now here he is again, in this Zoom room that feels incredibly small. I wonder if I could turn my camera off. The facilitator said not to unless it was an emergency. Does this count? I feel that to him, it would not count. So I keep the camera on, and I sit there. Silent. Awkward. Defensive.

I try some of the positive self-talk the facilitator lectured about earlier in the day. "Interrupting happens on Zoom," I tell myself. "It's not always intentional. It's not all about YOU."

I pick up my phone and text a friend who isn't in the session. "Have you ever had an issue with _____?" My friend texts back almost immediately. "YES. I almost went to HR about him last summer..." She goes on to provide a colourful description of this particular person, and how he's disrespected her in the past with his characteristic interrupting. She makes me smile, even though our gossip feels childish. Is childish. Anishinaabe Elders often advise us to avoid gossip. When we put negative thoughts out into the world, they go to that person. Negative thoughts can come back to us too. I know this, but the gossiping reassures me. It feels good.

Though I'm having a hard time reframing my mindset, I tell myself I should at least put my phone down and pay attention. I do. Another male manager is sharing a story. Despite this person going on at length, Interruptor just smiles and nods along. He does not unmute. Eventually, Interruptor, our self-appointed breakout room facilitator, asks "Would any of the faculty in the room like to share their perspectives?"

I wait. Nobody speaks. So, eventually, begrudgingly, I do. I take a deep breath and unmute. Picking at my cuticles off-screen, I start talking about the challenge of listening to so many students' stories, stories of suffering, of grief and loss, of real hardships that I don't feel equipped to assist them with. I explain how I find myself talking about these students during dinner time, thinking about them before I fall asleep at night. I'm surprised that I'm feeling a bit emotional, opening up in this way. The folks in the room are nodding.

"WELL, that's not your job to offer counselling sessions," Interruptor interrupts.

I see my own Zoom video projected back at me. My mouth is hanging open, mid-sentence. I let out a long, tired breath as he starts talking about on-campus resources and 1-800 numbers to deflect students to. Off-camera, one of my cuticles starts to bleed.

"Excuse me," I hear the words before I think them. Oh no. "Would you please let me finish talking?" I see Interruptor roll his Zoom eyes as he mutes himself.

I have so much I want to say, but my words have slipped away.

"It Finally Feels Safe"

I'm meeting with a student over Zoom. She's fallen behind in one of my online courses and wants to connect to get back on track. I've got the course and her grade report open on one screen, and Zoom open on another.

I smile at her when she joins. "Aanii! What's going on?"

We talk about the latest lockdown and what a rollercoaster it's all been. I notice she's quick to laugh. She has a childlike giggle that sounds almost like birdsong, even though I'm pretty sure she's my age or older. Her hair is dark with a strip of indigo blue running through it. There's a traditional painting hanging behind her as she sits on her couch. I comment on how beautiful it is, and she tells me it's actually a blanket. She takes it down and holds it up to the camera so I can get a better look.

She folds and refolds the blanket as she tells me this is her third time attempting college. I hear noises in the background, little voices, and she says, "One sec." She leaves her camera and mic on but disappears off-screen. I hear some chatter, some banging noises, and then she's back in the camera frame again. "Sorry! Little scuffle over a toy happening here!" she says, laughing, then continues telling me her story.

She shares that when she first left her home community things didn't go well for her and she fell into a rough crowd. "I flunked out," she says. "But I knew I could do better." I nod. We talk about how life paths are rarely ever straight lines. She nods then tells me about her kids, three beautiful Ojibwe Anishinaabe children who are currently all at home doing virtual school.

"They're listening to me right now," she says and laughs her sing-song giggle.

"I bet they love hearing your compliments," I say with a smile.

I think about my friends who are both teachers and parents, about the struggles they've been sharing on Facebook. The impossibilities of working full-time while parenting full-time, all the while doing their best to cope with the existential crises of the pandemic. I quip, "It must be hard for you to be a full-time student, parent, AND I.T. support!"

Her smile disappears. She pauses, then says, "Actually.... It's been so great having them all home with me. Their father is abusive. We left last year. Honestly, I've been so scared that he's going to steal them from school. It finally feels safe, having them all here with me..." she gestures off-screen, toward her kids who I imagine are laying on their tummies or sitting on tiny kid-sized chairs, maybe colouring or listening to their teachers through headphones.

She says quietly, almost to herself, "Kinda funny eh, to finally feel safe in the middle of a pandemic?"

Neither of us laugh.

"Just Woke Shit"

For the past several semesters, I've been designing and facilitating an asynchronous online course titled **Truth and Reconciliation**. In the course, students take a sustained, in-depth look at the residential school system in Canada. I introduce concepts such as settler colonialism, terra nullius, the Doctrine of Discovery, intergenerational trauma. We listen to stories shared by Survivors and their families. We watch archival footage. We learn about the Truth and Reconciliation Commission and critically examine the progress (lack of progress) on the 94 Calls to Action. We learn, reflect, research, share, celebrate the good work, criticize the lip service and assumptions.

Teaching this course is very humbling. At this point, I've put so much of myself into it, and through several iterations, with the guidance of many helpers, including hundreds of learners who've passed through, I've made improvements and feel it's in a good place, always though, with space to do better.

The final assessment in the course is a free writing (or free talking, should students wish to audio-record) exercise that invites students to reflect on their experiences, to think about what they've learned, and what they hope to remember long into the future. These submissions are often incredibly inspiring and energizing for me. Through their sharing, students tell me that the course is working, and I'm grateful for the reassurance. They also leave me with a lot to think about in terms of future improvements, additions, clarifications.

I was sitting in my little makeshift at-home office, reading and listening to these final reflections, sipping a tea, and feeling pretty good about myself as an educator. Then I came across a submission that popped off the screen and slapped me right across the face.

The submission basically read:

> I hated this course.
>
> What's the point of learning this stuff? Nothing can be changed. We can't right these wrongs, and anyway, it isn't our fault. All this course and stuff like it does is promote negativity. It's US versus THEM and this woke shit is tired. It's just creating more division from years gone by. People just love to point fingers. Then when it happens to them, they just do it right back.

Blink. Swallow. Deep breath in.

I wrote, "Meegwetch for sharing your perspectives" then slammed my laptop shut for the day.

"You're a Good Kwe"

I teach an online course that centers around critical thinking. Last term, I was loaded with a blended delivery, meaning I got to spend an hour with students synchronously, via Zoom, and the remainder of the course activities took place asynchronously, through our learning management system and other online spaces.

I wanted to problematize the Eurocentric assumptions underpinning "critical thinking skills," and so early in the semester, we started talking and learning about how worldview and culture impact our ways of knowing, being, and doing in the world. I approached one of our Elders on campus, Nokomis, and told her about the course and my goals. I offered semaa and asked if she would join us for a virtual guest lecture.

She agreed to share her teachings of Mino Bimaadiziwin, the ways of being and stories that help Anishinaabe people live a Good Life. I prepared my students for her visit several weeks before she joined us. It was important to me that the students in the course, who were primarily non-Indigenous, appreciate the Sacred nature of these teachings and principles. That they understood what it meant to be in ceremony with an Elder, even if that ceremony was taking place via Zoom.

One of the students in the course required all Zoom classes to be recorded, as per her official Accommodation from our on-campus Accessibility office. I told Nokomis about this, and she was fine with being recorded for this purpose.

Day of, Nokomis logged on, checked her mic, and immediately immersed us in her beautiful teachings. She smudged, prayed in Anishinaabemowin, sang a traditional song, and shared the principles of Mino Bimaadiziwin. She kindly invited student questions, which resulted in a wonderful exchange. She is so generous in sharing her wisdom. As the class wrapped, I felt warm and hopeful. "Chi-meegwetch, Tech-Enabled Nokomis!" I said as we were logging off. She laughed a big belly laugh, and that made me feel good. That night, I prepared some gifts for Nokomis. I brought them to the next Full Moon Ceremony and offered my gratitude. She gave me a big hug.

A few weeks later, a faculty member instant messaged me through Zoom chat. "Hey Jess," she said, "I heard Nokomis did a teaching for your class recently. Can I have the link to the recording?"

I started to type a response but hesitated. The hairs on the back of my neck woke up. I moved my fingers away from the keys. I didn't know how to answer.

Instead, I telephoned Nokomis, who was just about to head out for her daily walk. I explained the request, how it made me uncertain, and that I was calling her to make sure it was okay before I shared the link. My voice sounded sheepish. I felt flustered.

After a short pause, she said "No. You asked in a good way. You offered semaa. She didn't, and I don't wish for that recording to be shared in this way."

We talked for a little bit longer. At the end of the call she said, "It was right that you called me Jess. You're a good kwe."

I sucked in a deep breath as I hung up the phone. Nokomis thinks I'm a good kwe... I'm still contemplating what that means to this day.

"Wichita Do Ya"

Full-time employees at my college can take any for-credit course for $20, plus the cost of course materials. I asked a friend and faculty colleague if she'd be comfortable with me joining her course. "Sure, but do you have time for that?" she asked. "I'll make time," I said. "Well, I just hope it's worth your $20!" she joked.

She and I often talk at length about Indigenous education, about the challenges that come with teaching courses that some students are resistant to, the micro and macro aggressions, the imposter syndrome and constant feelings of self-doubt. We swap stories and resources daily. We attend webinars together, virtually, keeping up a lively backchannel and sharing recordings with each other when one can't make it. We've sat in ceremony together. "I heard your voice tonight," she said after our first Full Moon Ceremony. We've crafted together, paddled, hiked, swam, snacked. One time, we watched a mother bear and her three cubs grazing in a field not far from us. It was beautiful and spiritual, especially for my friend who is mukwa dodem.

As we learned about each other and got more comfortable, we shared stories about our families and our Algonquin heritage. There are intersections and overlaps between our stories, our friendship, our ancestors. At times, it feels like we were compelled by greater forces to find each other, to get to know each other, to learn together and lean on each other, just as our relatives did before us.

Now, I'm a student in her online course. I trust her, so in my course assessments, I get personal. I try to make deep connections between what she is teaching us and what my family has lived through. It deepens our friendship. The example she sets makes me a better educator.

On the very last day of the semester, my friend conducts a Closing Circle via Zoom. She's lit a smudge. I do too. It feels weird for me. Usually, I turn all of my electronics off to smudge, but I can roll with it. I turn my camera off so that I can ground myself and find some balance.

My friend explains how the end of the semester can feel like a rushing river, blowing us around in an out-of-control way. I think about this first pandemic year, the raw moments I've shared with students, my self-doubts and constant questioning. She explains how choppy waters eventually calm down. "Hold onto that knowledge," she says. I think about how grateful I am for my safety and health, for my family and friends who love me, for the chickadees that fed from my hand all winter, for all my relations. I remind myself to be grateful for a good job that is fulfilling and intellectually stimulating. I exhale and feel my body relaxing.

My friend picks up a ceremonial shaker and introduces the Water Song. She sings it for us.

Her voice is strong and steady.

I close my eyes and hear the harmony. Though I'm muted, I quietly sing along.

Intentionally Equitable Hospitality as Critical Instructional Design

Maha Bali and Mia Zamora

Intentionally Equitable Hospitality (IEH) is a facilitation praxis that was first developed by the co-directors and members of the grassroots movement called Virtually Connecting (Bali, Caines, Hogue, DeWaard & Friedrich, 2019). Virtually Connecting (VC) has "challenged academic gatekeeping via rendering private hallway conversations that build social capital at face-to-face conferences into public hybrid conversations in which people who cannot attend conferences are able to participate" (Bali et al, 2019, para. 7). As such, IEH was initially focused on hybrid professional development intended to promote equitable access to conversations, with multiple volunteers as facilitators. This paper recontextualizes the concepts and spirit behind IEH to work in a formal educational context, replacing facilitators of VC conversations with teachers in formal contexts, and replacing participants at conferences with students in classes.

IEH begins with the notion that the teacher or workshop facilitator is a "host" of a space, responsible for hospitality, and welcoming others into that space. IEH requires intentionality about who is involved in the design of that space, noticing for whom the space is hospitable and for whom it is not. IEH is iterative design, planning, and facilitation in the moment. It also includes the interactions outside of formal gatherings that influence formal, synchronous interactions.

As Priya Parker (2018) has suggested, the way we gather matters. This observation holds for educational contexts. A class is often a unique entity, with its own chemistry or "personality". It holds particular memories. A class occurs at a particular time in one's life, and it is experienced in a particular place. Learning together holds the potential for unique growth moments, and can be truly transformational if it is tied to a sense of belonging. If a student gains the experience of being included and heard, it makes a critical difference in what kind of learning is possible for all.

But this aspiration is often at odds with institutional mandates that hem teachers in with an emphasis on content and prescribed learning outcomes. How can teachers foster an authentic and collective sense of belonging when designing for impactful learning? How can they create an equitable environment that is hospitable to diverse students? Since all gatherings are essentially collective endeavors, learning design for equity-in-community is a critical component of IEH facilitation. IEH is a values-based approach that promotes co-learning among students, who might be different in innumerable ways, by prioritizing the needs and wants of the most marginalized among them.

As a critical pedagogical approach, IEH centers values rather than measurable predetermined outcomes. Intentionally equitable hospitality is not neutral. Rather, it prioritizes the values of social justice while fostering learner/participant agency within the learning space, while never forgetting the ways in which power and oppression work outside of that learning space, and how they influence it. For educators, the "intentionality" at the core of an IEH approach is a crucial first step. When we wish to practice IEH, we need to continually renew our intentions to notice oppression and injustice and seek to redress them, to iteratively modify and adapt our practices according to the responses and reactions of participants/learners, particularly those who bring marginalized perspectives.

When we say "we", we don't mean "we" educators only, but a practice of spreading an IEH mindset and praxis for everyone participating. The work requires a constant renewal of the daily effort to pay attention, to interrogate one's own positionality, to imagine and extend one's own critical engagement with "the other", and to reckon with the limits of one's own understanding of other people's lives—to model this and explicitly discuss this so that co-learners can begin to practice IEH with one another. As such, IEH works on multiple levels and in different directions, not all of which will work in harmony in every context—a behavior considered welcoming by one group may exclude another. Ultimately, the educator is host, but never the gatekeeper. The students are essential co-creators of meaning in the learning community. This seems a simple paradigm shift, but takes a significant amount of critical self-reflection.

IEH work is always in process, taking the form of an aspirational journey, but never an arrival. Educators must embrace iterative modification, emergence, and revision thinking. This article aims to set a framework for how a teacher, instructional designer, or faculty developer can incorporate IEH into their learning design, and it shares real stories from our practice.

Why Do We Practice IEH?

Learning environments, whether virtual, hybrid, or in-person, are inherently inequitable, for two reasons. First, because they mirror the outside world, encompassing the range of oppressions including white supremacy, heteropatriarchy, capitalism, colonialism, ableism, xenophobia—and all of the intersections within participant identities in the learning space. Second, because learning spaces introduce their own power dynamics, related to institutional cultures and how learners are expected to interact with authority figures, and related to hidden curricula that encourage or hinder certain behaviors, including, for example, host controls over muting/unmuting in video conference platforms, seating arrangements in in-person classrooms, and angling of cameras and location of microphones in Hyflex/Dual Delivery classrooms. Individuals within these learning spaces have an unequal opportunity to learn, to contribute to everyone else's learning, to grow. If we cannot control oppression and social injustice outside of the learning spaces we design, at the very least we can resist them within spaces we are able to control and influence, recognizing that our work will always be a journey, it will always be partial and that we continually strive for more, because:

> Marginality can be visible and invisible... Those at the centers can never see what it looks like to be on the margins, because the world looks different from the margins...
>
> This unique perspective, especially on suffering and in-betweenness, is why we need to have voices from the margins in our textbooks, teaching our classrooms, managing our educational institutions and representing us in media and in government. This is why Kamala Harris matters so much to girls, black women, Asians and immigrants in America. Because cognitively understanding systemic inequalities does not prepare you to truly understand the experiences of being marginal/ized (Bali, 2021, para. 9-10).

Representational justice matters in the classroom. We have a responsibility to elevate voices historically oppressed by privilege. IEH includes the ongoing work of representational justice—to understand that representation matters and that we all possess the agency to manifest narratives of both our neighbors and ourselves that embody humanity, truth, and respect. Attending to representation in a classroom can be corrective—undoing past misrepresentation—or reflective—embracing the full breadth of complex identities existing around us. When we create equitably hospitable spaces, we help nurture the growth of everyone present, so that each co-learner may gain more insight and more power beyond the boundaries of the spaces we traditionally inhabit.

The work of "opening up" conversation to diverse lived experience in a classroom context is an essential component of the social justice values underpinning IEH. When students can discover stories "that feel like home" in their curriculum, or even better, if they find the courage to share stories that represent their own perspective, we see the beginnings of trust-in-community. Learners honor the stories of others— whether empathizing with those stories comes naturally due to shared experience or uncomfortably—via epistemic listening to stories from perspectives one has never encountered fully before. This intimate and deliberate design work allows for narrative emergence in classroom spaces and aspires to create the conditions for co-learning in compassion. But this kind of trust can only be built with careful and equitable forms of engagement. IEH is a map for this nuanced and ongoing effort.

When Do We Practice IEH?

One of the main criticisms of traditional instructional design is its over-emphasis on design ahead of when actual teaching takes place. It assumes that if you design something thoughtfully enough, with some sort of empathetic imagination as to who students are and what they may want, then when you enact this design in practice as a teacher, everything will fall into place. However, what actually happens in practice may differ from the plan (see Alhadad, Bali, Gachago, Pallitt & Zamora, personal communication; Bali, M., Gachago, D., and Pallitt, N., 2022;)

IEH does not dismiss the importance of planning and design, but it recognizes that making a space hospitable happens at various times in the process:

1. **Pre-design:** Whom do we consult in the process of imagining our design? Do we contact past students? Do we discuss with colleagues from diverse, especially marginalized, backgrounds? Do we have access to our current students and how far can we involve them in the design?

2. **Design:** In what ways can our design anticipate power inequities, and in what ways can these be redressed a priori? In what ways can we leverage flexibility and openness in

design and diverse pathways and "Universal Design for Learning" in order to make room for participant/student voices to modify the design, a design that is inherently participatory and emergent with learner agency baked into it by design?

3. **In the facilitation/teaching moment:** What kind of practices promote psychological safety for diverse participants to participate fully and also speak up when something is not comfortable or welcome? Can we move away from safe to brave spaces? How do facilitators respond when something goes wrong, when one participant's behavior may create discomfort, offend, or actively oppress? How do participants themselves respond?

4. **Beyond the moment (sustaining community):** How do we iterate, beyond a particular teaching moment to reflect on our own, and with our students/participants in order to help us notice anything we may have missed, and imagine how to do better next time?

Some of the practices of IEH as done in VC that transfer directly to classroom teaching include:

- Prioritizing listening to the more marginalized voices and ensuring everyone has an opportunity to express themselves.
- In the same way that VC invited participants to select conferences and guests for sessions, students can be asked (e.g. via surveys or in dialogue) to choose topics for the day, for example, ahead of class or during class.
- Creating spaces where participants, rather than facilitators, can choose and change the direction of the conversation, embracing serendipity.
- Ensuring participants have agency to participate in the ways that feel most comfortable for them, via audio or chat, with cameras on or off.
- In recorded conversations, having time before and after a session that is less formal and unrecorded, where some people may feel more comfortable speaking and interacting.
- Using semi-synchronous spaces to sustain community between synchronous sessions.

How Do We Notice Social Injustice and Oppression?

Social injustices are multidimensional. One typology of social justice is Nancy Fraser's (2005), which differentiates between:

- Economic injustice, such as access to digital devices and infrastructure, which can be redressed via redistribution or resources;
- Cultural injustice, which involves the erasure or misappropriation of certain cultures, such as the absence of Indigenous and non-white cultural perspectives from many Western curricula, and can be redressed via reappropriation; and
- Political injustice, which involves the inability to participate in democratic decision-making about one's own fate and circumstances, and can be redressed via participatory parity.

These three dimensions are, of course, not mutually exclusive; people's identities are intersectional. For example, many populations suffer from a combination of these, such as African Americans. In some contexts, a group may suffer one kind of oppression but not in others. For example, Christians are the majority in many countries but are religious and cultural minorities in Egypt, where many of them are economically prosperous but overall their culture is not well-represented in curricula. In this context, they may therefore suffer ideological and interpersonal oppression. Economic inequality becomes particularly important in online learning if sessions are synchronous and students do not have access to dedicated devices, stable internet bandwidth, or private spaces to learn in their homes.

Oppression exists on many levels, from ideological oppression in society (racism, sexism, etc.), to institutionalized oppression (policies and laws that are unjust, such as redlining), to interpersonal oppression (such as microaggression) and internalized oppression (such as impostor syndrome). Recognizing these many levels and naming them, and how they manifest within our learning spaces can help us attempt to redress them. When we, those with power to control and design a learning space, recognize these oppressions, it enables us to promote "parity of participation" (Fraser 2005) for the learners in the design of the learning process.

Inviting everyone to participate in creating "community participation guidelines" as a democratic process does not remove the power dynamics amongst the participants or the internalized oppression of some participants. This may leave them silent or unable to object if other, louder voices, suggest items they do not agree with. Indeed, internalized oppression may mean that those unused to having agency may make decisions harmful to themselves, for lack of imagination of their own potential or of alternative options (Walker & Unterhalter, 2007). For example, students unused to having agency might, when given the choice of presenting their work as a regular paper or using any multimedia of their choice, fall back on what they have internalized as considered "good academic work" traditionally. How do we create environments that dismantle oppression, while raising critical consciousness of all involved, in order to achieve "parity of participation" so that the most marginalized in our group feel empowered as co-owners of the space?

> "Feminist education for critical consciousness is rooted in the assumption that knowledge and critical thought done in the classroom should inform our habits of being and ways of living outside the classroom" (hooks, 1994, p. 94).

Stories from the Field: IEH in Practice

We have critiqued instructional design that assumes a good design done ahead of time will produce particular results. In reality, much of the work of creating socially just learning spaces requires facilitation in the moment, whether this is done via negotiation amongst learners or by the facilitator listening and responding to learners. Centering equity in the design process helps reduce equity emergencies (e.g. one person dominating a conversation) but does not eliminate them. Similarly, a well-facilitated session centering equity is incomplete without pre-work that promotes equity (e.g. who is involved in designing the session, how is the time agreed upon, is there community building done ahead of time?) and post-work that builds on the session to take equity work forward.

The examples we offer of IEH applied in educational settings may not apply IEH in all the ways VC has done so, but show its spirit in various ways.

Virtually Connecting

IEH originated from how the VC community strove towards equitable hospitality before, during, and beyond conversations. Before a conversation took place, work was done to ensure those furthest from justice could choose which events to do VC at and select guests at conferences with whom they would like to speak. This work involved inviting guests and ensuring representation of women and marginalized groups in the guest lineup where possible. It also involved announcing sessions and personally reaching out to virtual participants to join sessions when they showed interest but didn't register. During the session, the presence of both onsite and virtual buddies ensured there were hosts welcoming people in and checking that needs of both sets of participants were met, and participants could focus on the conversation while technology was taken care of. Beyond the session, VC built community such that IEH was a central value and praxis we developed, and we often reflected asynchronously or synchronously after events on ways we could improve the equity in our hospitality. Beyond learning how to manage the technology to run a VC session, new volunteer buddies shadowed more experienced buddies to learn from the ways different people modeled equitable hospitality to fit their personalities. Occasionally, VC would solicit feedback from participants formally, but informal feedback was collected all the time.

What hyflex classrooms can learn from Virtually Connecting

A model of hybrid synchronous teaching called Hyflex often involves a teacher in a classroom, teaching some students in-person while others join virtually via video conferencing. This is a model that is both technically expensive and pedagogically complex to apply well, but some educational institutions attempted versions of it during the COVID-19 pandemic, as a way to ensure distancing and flexibility within classrooms without being fully online. This model has so much in common with the Virtually Connecting format, of having a group of people together in one space connecting with a virtual group. (Bali, 2020)

If IEH is not applied in a Hyflex setting, virtual students are likely to feel like second-class citizens. One way to counter that, to ensure virtual students' voices are heard and needs are met, is to have someone like the VC "onsite buddy," someone sitting in class who is responsible for checking in with virtual folks to check if they have questions or concerns. In cases where teachers do not have TAs, rotational student volunteers can do this role. There can also be something like a VC "virtual buddy" responsible for virtual students, helping with things like breakout rooms, etc., and perhaps communicating with the onsite buddy, or if they are a TA, occasionally facilitating smaller virtual conversations.

Without this kind of support, the teacher's attention will either be divided between the two groups, giving neither of them sufficient attention, or they will end up focusing on one group more than the other. Occasionally, equity-minded teachers end up focusing on the virtual students, which is still problematic for the in-person students.

As with VC, some degree of digital literacies and competence helps promote a more equitable experience. For example, when teachers want to solicit responses from all students, they can choose tools that both onsite and virtual students can use at the same time, using external polling tools rather than the ones within the web conferencing tool, or they can use collaboratively edited documents instead of writing on paper or the in-class whiteboard.

In all of this, listening to student feedback continually can ensure teachers redress any inequities that occur, and perhaps sometimes students themselves can suggest solutions to increase equity in a Hyflex classroom.

Reflections of IEH approaches in the classroom

Telling small stories to build trust

How can an educator lay the groundwork for a learning community built on sincere connection and trust? How can this be achieved when a group of different people comes together for a class? How can an educator convene a group of strangers and effectively foster both compassion and critical acumen?

You cannot insist upon trust. It has to be something that emerges from moments—moments that build upon each other over time. A formal class affords this time in ways one-off PD events do not. There are many quick warm-ups and introductory activities to engage students in IEH (see Bali, Caines & Zamora, undated). These moments make a big difference.

Getting to know one's peers is not necessarily an outcome that bears much academic significance in a traditional class. But to know your co-learners, little by little over time, is a key reason why most people feel motivated and connected in a course. Authentic and voluntary participation is also key to the success of peer learning modalities. I, Mia, incorporate many warm-up and reflective activities leading to short but sincere moments of sharing during class time. They are never a waste of time. Rather, these activities become a critical building block of IEH.

Small stories make all the difference in building a sense of community. There is secret power in a story. Stories can be a small gift given with purposeful intention. They can provide insight into the life of the storyteller. They are like a seed that is germinated (or a sachet of tea that slowly infuses in boiled water)—pervading gently in one consciousness, growing an understanding of something new. A small story can be a bridge between different people's perspectives.

In lieu of the typical introductory protocol of going around the room (or Zoom grid) to introduce oneself by name and major (a ritual that is at best boring, and induces dread and stress for many students), strategies like the IEH "Image Gallery" warm-up can be used (adapted from "Four Ideas for Checking In"). Here, everyone looks at a grid of randomly selected abstract images. Examples of pictures in the gallery might include a cozy fireplace, a bright electric cityscape, a rugged mountain terrain, or a labyrinth garden. Each co-learner is asked to choose an image from the gallery and explain why they chose it. How does it relate to how they might feel at the moment? What does that image mean to them and why? The result is numerous small stories that each individual chooses to share. The critical element of IEH design here is that each participant has the agency to choose whatever insight they want to share. There is no expectation or prescription of what one must tell others about themselves. It is the interpretive openness of the prompt—"What do YOU see?"—that generates something unique from each participant. The insights yielded are the glue of growing trust and understanding between co-learners.

Student remix as IEH co-creation

As a teacher-facilitator, I, Mia, am always exhilarated when I see signs of IEH values taking hold. IEH is evident when students intuitively suggest ideas for new forms of co-creation. It happens when students come up with their own (unsolicited) ideas for how to actively engage with each other. By imagining new ways to remix activities, they build upon IEH foundations and make evident new levels of trust.

A simple example of IEH student re-mix was when a thoughtful student asked to return to the Image Gallery protocol towards the close of a course. The special twist on the original protocol was that instead of each student selecting an image and then sharing a personal insight, this time each student would select an image inspired by thinking about another student in the course. What image in the gallery resonates when thinking about a peer in the course?

In this case, I immediately knew that it would only work as an equitable protocol if each and every student received a "shout out" from another student. In the class of 20 students, I designated 5 random groups of 4, and asked each group to self-assign who they would select an image for within that small group, so that it was ensured that every person was accounted for. Since this remix was conducted in Zoom, I also invited them to send "shout outs" to other students through use of the chat if they wanted to. These would be "extra" insights in addition to the one "shout out" they shared with the overall group. I knew that many wanted to share their positive insights with more than just one other student in the class. The result was an exhibition of generosity, insight, and complex forms of trust. Students shared things they noticed in each other, and reminded everyone of knowledge they gained from certain individuals along the way. They paid particular forms of tribute to each other, and they made evident how each person has their own unique way of contributing to a community's overall learning.

Liberating structures in faculty development workshops and classes

I, Maha, am a faculty developer, and when the COVID-19 pandemic forced us to move everything online, I tried to apply IEH in online workshops, to model it for faculty, so they could use it with their students. I quickly realized that VC-style conversations without topics and informal dialogue, are difficult to facilitate equitably in larger groups, say, larger than 12 or 15 (something Autumm Caines, 2015, often calls "The Interpersonal Multitudes Barrier").

When we offered "morning coffee" and "ask us anything" sessions that were announced as open dialogue, centering faculty members' interests and needs over faculty developer agendas, this usually worked out because numbers were small enough for everyone to participate,

21

like a VC session. However, with larger, more formal workshops, we needed something different.

At first, we used polling tools, the chat, and Google Docs, but this still centered the workshop facilitator's agenda and privileged text-based over oral interaction. It seemed better to use breakout rooms, have people discuss in smaller groups, and then report back. This potentially gives time for every voice to participate and for a workshop to feel more interactive. However, poorly designed breakout sessions could be disastrous: you had to balance the number of participants per room, the timing, and the breakout settings. It was sometimes necessary to have a facilitator in each breakout room, which was not always logistically feasible, and we also learned that breakout groups needed very clear instructions to follow or they felt lost or awkward. This is where virtual Liberating Structures (LS) were a saving grace.

Liberating Structures are "are easy-to-learn microstructures that enhance relational coordination and trust. They quickly foster lively participation in groups of any size, making it possible to truly include and unleash everyone... [They] can replace more controlling or constraining approaches." Among the ten principles underlying the design of LS, the two most relevant for IEH are:

1. Include and Unleash Everyone
2. Practice Deep Respect for People and Local Solutions

When we used LS in our workshops, faculty members' key takeaway was often to plan to use more breakout rooms in their classes, and their key PD need was to learn how to use breakout rooms smoothly. In response, we offered hands-on workshops on creating breakout rooms and workshops explicitly teaching LS. Students since then generally gave positive feedback on the use of breakout rooms in classes.

LS are designed to promote equitable, respectful collaboration, through the sequence of steps, distribution of timing and group size. Originally designed for in-person collaboration, most LS work well with virtual breakout rooms.

Three particular structures demonstrate IEH by respecting all voices equitably:

    A. **1-2-4-all** (https://www.liberatingstructures.com/1-1-2-4-all/) is a dialogue version of "think-pair-share". In 1-2-4-all, a facilitator shares a prompt. Each participant thinks about it on their own first. This gives people who are more reflective, non-native speakers and marginalized groups a chance to think quietly without interruption and formulate their thoughts before sharing in the second stage of sharing with pairs. The next stage is discussion in a group of four before sharing with the larger group. When people test out their ideas in small groups, they all contribute and gain confidence while building on each others' ideas before sharing with the larger group. It ensures everyone participates, in contrast to most large group discussions, where certain individuals tend to dominate the discussion.

B. **Conversation Café** (https://www.liberatingstructures.com/17-conversation-cafe/) places participants in groups of 4-5 to respond to a prompt, giving micro-timing for individual participation within rounds (and thus it helps if someone volunteers to be the "timer"). In the first round, each person takes one minute to respond to the prompt, and in the second round, each person responds to other people's responses in one minute. Afterward, there is some open conversation time before wrapping up with key takeaways that a "volunteer note taker" writes on a shared document. This approach approximates equity by giving equal time, but of course, some people need time to think before speaking (and would benefit from 1-2-4-all). Others might need different amounts of time to express their ideas, either because they are non-native speakers, they take time to express complex ideas, or they just speak more slowly. IEH might ask people who speak more quickly to give up their time for others, give people time to write quietly in a shared space before sharing orally, explicitly give more time to marginalized groups (e.g. students in a student-faculty conversation), or let marginalized groups speak first.

C. **Troika Consulting** (https://www.liberatingstructures.com/8-troika-consulting/) is another LS that uses microtiming and introduces reciprocity well. Each group of 3 works together in 3 rounds, switching roles each round. Every person has the opportunity to act as a "client," asking for consultation from others, and the other two perform a quick consultation using a specific format; all of this occurs within 10 minutes. This structure is surprising in how quickly, in the space of 10 minutes, one can receive useful ideas for a challenge one is facing, from people who may not be experts. The reciprocity of asking for help once and giving help twice in the same structure tends to also be satisfying for participants. This kind of structure done within faculty development workshops empowers faculty to set the agenda for what they want to discuss, to seek help from peers rather than faculty development experts, and when used in class, encourages students to seek help from colleagues not just look to teachers for help. It may help people to be given time to prepare for their "challenge" ahead of time before they go into groups since the time is usually limited and the process quick. It is important for facilitators to create a relaxed atmosphere before sending people into their groups so as not to create anxiety over the time limits.

For any LS, whether participants are strangers or not, it helps to dedicate time for people to introduce themselves or say hello before the activity. This alleviates the stress of working on the task, which could be difficult for some people to do without a warm-up.

A limitation of LS is that they require participants to "buy into" the participatory approach. If someone with lots of power decides not to follow the process, they may not end up making room for others to speak. Moreover, following a process that listens and builds upon ideas of all voices in a one or two hour workshop does not mean this will automatically transfer into changing the culture in a work environment unless there is further intentionality.

Some ways of addressing this is to use the facilitator's power with what Priya Parker calls "generous authority". Parker writes: "A gathering run on generous authority is run with a strong, confident hand, but it is run selflessly, for the sake of others. Generous authority is imposing in a way that serves your guests. (Parker 2018, p. 81). Replace "gathering" with "learning space" and "guests" with "learners/participants". Generous authority" is using power to achieve outcomes that are generous, that are for others" (Parker 2018, p. 82). It involves protecting participants from others who may hijack the experience for their own agenda, temporarily equalizing participants despite hierarchies outside the space, and helping participants connect with one another (Parker, 2018).

In practice, when using LS, the facilitator or teacher needs to explain the process to everyone and perhaps make explicit ground rules around sharing space and having all voices heard. If the groups are extremely unfamiliar with such approaches, there may need to be some co-facilitators present with small groups or moving between groups to check in and help out. The facilitator is likely to succeed better if they use inclusive warm-up activities that allow participants to get to know one another and get into the mood with low-stakes activities and discussions that do not build on anyone's authority. Some participants will resist unless they know the purpose of every step. It is sometimes worthwhile to stop and make time to discuss purpose before or after an activity, to help more participants stay on board.

The facilitator's generous authority involves reminding people of timing—its importance for equitable and productive dialogue—and designing who ends up in which group with thoughtfulness. For example, if the facilitator knows the audience well, they may avoid placing extremely dominant participants with extremely shy ones, unless they can have a co-facilitator to support equity in that group. The facilitator may decide to ensure marginalized groups are not tokenized in groups, e.g. never have one student among five faculty, but rather 2-3 students with two faculty. There also needs to be a safe way for a participant to report issues or seek help from the facilitator at any point. Finally, participants unwilling to try these ways of doing things may choose to "pass" but not be "counted" among participants, possibly observing, but indicating clearly so facilitators don't count them among active group members in something like Troika. It is important to recognize that people who resist these approaches may come from all walks of life: they may be very powerful people unwilling to give up on hierarchy, or they may be marginalized groups unwilling to make themselves vulnerable. It is important to remind ourselves of the purpose of a gathering, invite the right combination of people first, and design the groupings and the flow of the experience in ways that help us meet the purpose; it is also important to remind ourselves these decisions can be really complex (Parker, 2018).

In terms of follow-up beyond a session, it would depend on the facilitator's relationship with and power amongst the group of participants, but at the very least, a live session or meeting can end with agreements on future commitments or next steps based on the dialogue. Where there is a sustained relationship among facilitators and participants, there may be some degree of negotiation behind the scenes before and after sessions that can help make sessions more constructive and fruitful beyond the meetings themselves. Reflecting on higher education, the power dynamics when a faculty developer is offering a workshop to tenured professors in one department is different than if they were a learning community across disciplines that meets regularly and is very different from a class. And a class of 15 in a seminar-style classroom is different from a class of 40 learning online which is different from a class of 80 learning in a large lecture hall.

Community-Building resources

In August 2020, we (Maha, Mia) collaborated with Autumm Caines and approached an organization called OneHE (whose goal has been to enhance higher education globally) to suggest a collaboration with "Equity Unbound" (Equity Unbound is an emergent, collaborative network which aims to create equity-focused, open, connected, intercultural learning experiences across classes, countries, and contexts). The collaboration we proposed was to create and curate an Open Educational Resource that demonstrated to educators worldwide a variety of ways for building community online. We felt this was needed because people unfamiliar with online teaching needed ideas for building community online, especially in these traumatic times of physical distancing. Faculty developers and educators worldwide were overwhelmed, so we designed these resources with demonstrations and templates they could adapt.

These resources practiced IEH by:

1. Involving multiple people around the world from different contexts in their design, so people who experienced online differently or had different resources could notice and point out inequities or challenges with each suggested strategy for community building

2. Offering adaptations for elements that may not be accessible to some groups. E.g. synchronous and asynchronous options; breakout and no-breakout adaptations

3. Inviting feedback from the public and inviting viewers to contribute. We used critical and constructive feedback on the resources, to iterate and adapt them since.

For concrete, practical examples, see Bali & Zamora (forthcoming, in *Learning Design Voices*)

IEH in Equity Unbound studio visits

"Studio Visits" are a way to enrich both classrooms and professional development engagements with the practical experience of experts in the field. The visits are virtual by nature, and bring together diverse co-learners from different locales, and with different levels of training or professionalization. A studio visit is marked by its unscripted nature, and there is little to no formal agenda set for the time allotted, as people enter into fluid conversation around a shared interest. A Studio Visit encourages co-learners to have informal, open discussions, giving participants access to dig deeper into how someone's scholarship, research, art, or practice has developed. The Studio Visit inherently challenges the hierarchical power dynamic of experts "presenting to" students as authority figures. We often host Studio Visits on a range of topics on behalf of "Equity Unbound"—an equity-focused, open, connected, intercultural curriculum that builds critical digital literacies in a global context, highlighting issues of web representation, digital colonialism, and safety/security risks.

Sometimes Studio Visits are conducted within a closed class setting with participation limited to the professor, the guest speaker, and the students. In these cases, IEH facilitation is a matter of extending a pre-established classroom culture. At other times, Equity Unbound Studio Visits are open. Professors join students from varying countries and cultural contexts, and co-learners represent diverse institutions. In these cases, there is an explicit power difference between the participants. Educators engage with confidence and professionalism, often rendering some students unsure. Some may intuitively take a back seat to those who they perceive as more "qualified" to participate. Also, some faculty participants have established collegial connections with other faculty participants. Therefore, a Studio Visit can be an occasion to have a chance to virtually reconnect with colleagues. But acquaintanceship among a few may leave others feeling like inherent "outsiders." In these open contexts, we have discovered that IEH is crucial in pursuing equitable co-learning. We must be intentional in the way we welcome everyone. The design for "opening up" the Q & A time should prioritize students and unaffiliated co-learners to share insights and ask questions. We must also remind all participants to be mindful of the varying levels of professional training and to make space for everyone to feel included in the ensuing conversation. Without clearly stating these intentions, the conversations are likely to be "hijacked" by collegial senior faculty who unintentionally get carried away in their own participation, leaving little room for the less practiced.

Student-Faculty co-design sessions

I, Maha, recently co-facilitated a session where students, faculty, and staff worked together to brainstorm and prototype ways of integrating gender into the curriculum at my institution. Although we used participatory approaches from a combination of Design Thinking and Liberating Structures, and kept people in small, diverse groups to make time for every voice to speak, we discovered later that in some breakout groups, student voices were not heard because some of the faculty spent a lot of time talking and no one took on the responsibility to ensure all voices were heard. We had used another Liberating structure known as "TRIZ" (a Liberating Structure that involves solving complex problems in a playful way by going backwards, working towards an anti-goal) in that session, which does not micro-time the group time, versus something like Conversation Café, which does remind us to give everyone a turn and equal time.

IEH as a community endeavor: Beyond the facilitator

What the above two experiences highlight for us is that for IEH to truly work, it is not enough for a facilitator to have these values and enact these practices. It is important for the community to have internalized the values or at least a few key people willing to co-facilitate in small groups or as backup in big groups. This is much harder to achieve in one-off meetings versus ones where the same group meets multiple times, like a class, and can learn from the facilitator's modeling. Beyond internalizing the values, participants willing to help facilitate sessions need to practice the ability to recognize oppression and microaggression as it happens, have strategies for countering it, and have a strong enough voice to speak up against it and challenge it, even if it is uncomfortable or against someone with more power than themselves. This is a big ask, and may not be "safe" to do in every context. The original Virtually Connecting context usually involved people with different degrees and types of power but who did not work together in the same institution and have those formal hierarchies of power between them. However, within the context of an institution, such power dynamics cannot be overlooked. Therefore, it seems that in an institutional context or in one-off meetings, it may be necessary to ensure that those with the most power are reminded orally of the importance of IEH, and those with the least power are aware of their rights as well, so they may advocate for themselves. Having some co-facilitators there, especially when people are unused to participatory approaches, can help, because, in virtual environments. the facilitator cannot see the dynamics of all the rooms at a glance the way they would in a physical setting.

Conclusion

IEH may not come naturally to all educators and instructional designers. It requires the practices of both equity and care to work in harmony, where equitable pre-planning and design are embodied with care in the facilitation of a learning space (Bali & Zamora, 2020 & 2022). It is also important to emphasize that IEH is not one practice or a list of practices, because "the notion that one model of care will work for everyone is absurd...humans vary in their abilities to give and receive care" (White & Tronto, 2004, p. 450). Nel Noddings (2012) asserts that a caring approach means we "do unto others as **they** [emphasis added] would have done unto them." This may mean anticipating others' wants and needs and designing to address them, or it may mean a more democratic and compassionate approach that is more participatory (see Bali et al., 2022; Alhadad, Bali, Gachago, Pallitt & Zamora, personal communication).

IEH is a praxis that strives towards socially just care. In a "caring-with" democracy, we can set a goal of structuring institutions and practices so that each person's individual preferences can be honored." (Tronto, 2015, p. 34). Bali & Zamora (2022) name "socially just care" as follows:

> This is when social justice is realized and embodied in a caring manner by participants in a social space... We did not name it "democratic care" (after Tronto, 2015) because democratic processes do not necessarily lead to equitable outcomes. We did not call it "parity of participation" because such a term does not emphasize the importance of care in order to create the social justice end goal. Socially just care, rather, promotes social justice and parity of participation in its designs and planned processes, and is enacted with care such that it always iterates to nurture self-determination, agency and justice for all involved, in whatever manner meets their diverse care needs, and addresses the multiple dimensions of injustice individuals and groups may face. It distributes the care responsibility so that the care is not "partial," and it goes beyond the "contractual" equity to ensure it goes beyond words and documents and becomes the lived experience within a social space.

References

About. (2018, October 31). *Equity Unbound*. http://unboundeq.creativitycourse.org/about/

Bali, M. (2020, November 18). What hyflex livestream models can learn from @VConnecting. *Reflecting Allowed*. https://blog.mahabali.me/educational-technology-2/what-hyflex-livestream-models-can-learn-from-vconnecting/

Bali, M. (2021). Foreword. *Voices of practice: Narrative scholarship from the margins*. Hybrid Pedagogy. *https://voicesofpractice.pressbooks.com/*

Bali, M., Caines, A., DeWaard, H. , Hogue R. J., & Friedrich. C. (2019). Intentionally equitable hospitality in hybrid video dialogue: The context of Virtually Connecting. *eLearn Magazine*. https://elearnmag.acm.org/archive.cfm?aid=3331173

Bali, M., Caines, A., & Zamora, M. (undated). *Equity Unbound & OneHE Community Building Resources*. https://onehe.org/equity-unbound

Bali, M., Gachago, D., and Pallitt, N. (2022). Compassionate learning design as a critical approach to instructional design. In Burtis, M., Jhangiani, S., Quinn, J. (Eds.), *Toward a critical instructional design*. Hybrid Pedagogy. https://criticalinstructionaldesign.pressbooks.com/

Bali, M. & Zamora, M (2020). Four ideas for checking in. *Equity Unbound & OneHE Community Building Resources*. https://onehe.org/eu-activity/four-ideas-for-checking-in/

Bali, M. & Zamora, M. (2021, February 25). *#OpenEd20 Day 2 Plenary: Maha Bali & Mia Zamora*. Open Education Conference 2020. Virtual. https://www.youtube.com/watch?v=NEeZvM6_8UE

Bali, M., & Zamora, M. (2022). Equity/Care matrix: Theory and practice. *Italian Journal of Educational Technology*. https://doi.org/10.17471/2499-4324/1241

Bali, M. & Zamora, M. (forthcoming). Designing and Adapting for Community with Intentionally Equitable Hospitality. In T. Jaffar, S. Govender, & L. Czerniewicz (Eds). *Learning design voices: Perspectives from the margins*. https://learningdesignvoices.com/

Caines, A. (2015, December 24). A call for more #HumanMOOC discussion groups. Or. The very human problem of access with more thoughts on the Interpersonal Multitudes Barrier (IMB). *Is a Liminal Space*. http://autumm.edtech.fm/2015/12/13/a-call-for-more-humanmooc-discussion-groups-or-the-very-human-problem-of-access-with-more-thoughts-on-the-interpersonal-multitudes-barrier-imb/

Fraser, N. (2005, December 1). Reframing justice in a globalizing world. *New Left Review*. https://newleftreview.org/issues/ii36/articles/nancy-fraser-reframing-justice-in-a-globalizing-world

hooks, bell. (1994). *Teaching to transgress: Education as the practice of freedom*. Routledge.

McCandless, K. H. L. & Lipanowicz, H. (2022a). 1-2-4-All. *Liberating Structures*. https://www.liberatingstructures.com/1-1-2-4-all/

McCandless, K. H. L. & Lipanowicz, H. (2022b). Conversation café. *Liberating Structures*. https://www.liberatingstructures.com/17-conversation-cafe/

McCandless, K. H. L. & Lipanowicz, H. (2022). Introduction. *Liberating Structures*. https://www.liberatingstructures.com

McCandless, K. H. L. & Lipanowicz, H. (2022). TRIZ. *Liberating Structures*. https://www.liberatingstructures.com/6-making-space-with-triz

McCandless, K. H. L. & Lipanowicz, H. (2022b). Troika consulting. *Liberating Structures*. https://www.liberatingstructures.com/8-troika-consulting/

Noddings, N. (2012). The language of care ethics. *Knowledge Quest,* 40(5), 52-56.

OneHE (2022). *OneHE: Develop your teaching your way*. https://onehe.org

— (2020). *Warm up activities*. https://onehe.org/equity-unbound/warm-up-activities/

Parker, P. (2018). *The art of gathering: How we meet and why it matters*. Riverhead Books.

Tronto, J. C. (2016). *Who cares?: How to reshape a democratic politics*. Cornell Selects.

Virtually Connecting (2022). *Virtually Connecting*. https://virtuallyconnecting.org

Walker, M., Unterhalter, E. (2007). The capability approach: Its potential for work in education. In Walker, M., Unterhalter, E. (Eds.), *Amartya Sen's Capability Approach and Social Justice in Education* (pp. 1-18). Palgrave. https://doi.org/10.1057/9780230604810_1

White, J. A. & Tronto, J. C. (2004). Political practices of care: Needs and rights. *Ratio Juris*, 17 (4), 425-453.

Humanizing Online Learning
Creating connection, designing for care

Mandi Singleton and Nicolas Pares

"If the structure does not permit dialogue the structure must be changed"

–Paulo Freire

Through the pandemic, Mandi and I (Nicolas) have been having long Zoom calls discussing the challenges and issues of online learning. Frankly, we felt that none of these challenges or issues were new to us. As faculty developers and instructional designers at an online college, we felt like we had seen it all. While supporting hundreds of well-structured and routine online courses, we recognized the need to focus on what makes students succeed online more than just good design.

Our conversations moved beyond Bloom's Taxonomy, instructional video creation, and feedback loops and into the spaces of critical instructional design and humanizing online learning. As we watched courses with perfectly developed outcomes, activity scaffolding, and assessment experience extreme differences in student satisfaction and success, we began to wonder.

Online learning and advancements in technology have created less space for human connection – dehumanizing learning. Too much focus is thrown upon the rote user interface interactions and not upon how the students will feel. When the tools are structured by design best practice, how can we design and facilitate for connection and care? This drove us to reflect deeply upon our own teaching experiences.

Both of us were instructors and teachers before we were instructional designers and faculty coaches, and still teaching today. Our lived experiences and our teaching experiences and contexts vary greatly so our conversations and stories tend to be divergent. Yet when talking about successful online learning or hybrid online learning, we began to see themes of humanizing learning rise up in our observations and personal experiences.

Mandi comes from the K-12 space with an emphasis on STEM and the sciences. Having worked as an instructional coach in urban and title one schools and now a faculty member in a Bachelor's completion program, her experiences encompass various educational roles.

Nicolas's teaching experiences traverse the English language teaching space from adult education and K-12 along with some mathematics teaching in secondary high school, and now applied linguistics and TESOL teacher education in higher education.

These discussions from early 2020, raised a major question for us. How can instructors and instructional designers humanize online learning through pedagogies of care and designing for connection? To answer that we thought back on our teaching experiences and looked for what humanized our experiences and what humanized our students on their path to student success and meaningful learning.

Humanizing Online Learning

Before we could get to the initial question, our discussions took us through stories across a broad spectrum of teaching contexts where our students and we as instructors felt human. As the pandemic began to separate and create a distance that isolated learners and teachers, our virtual exchanges began to become more personal and focused on connection. Our coaching and instructional design work became more frequently focused on compassion and understanding towards students and each other. This led us to discuss stories of teaching practices and applied instructional design norms that we felt could empower connection and lead to student success and achievement. Connection and care became central to our thinking and teaching.

Some of these stories of humanizing and dehumanizing online learning are below. We used these stories to explore the question, "What does the experience of humanized learning feel like for instructors and students?" From these stories themes emerged. Those themes brought learner, instructor, and learning together, and bridged the gaps of educational systems and instructional design norms focused solely on single aspects of teaching and not the holistic human condition.

Ultimately, these stories and themes highlight the need to humanize learning but to go beyond the old, "human-centered learning" or common user experience philosophies that inform the rote, redundant course designs that are now so prevalent in higher education. To humanize learning is to foster a sense of care in your students and truly center human connection in design.

While the modern tech tools used to deliver online learning provide space for interaction via discussion forums reminiscent of the AOL chat boards or Tik Tok-esque video exchanges, we can not as designers and instructors rely upon the mere functionality of exchange to develop connection or care. We must be critical instructional designers and reach for a caring, connected, and humanized online learning experience.

Pedagogy of Care

Students often have many concerns when they enter a new learning space such as an advanced degree program with learning centered online. Will I feel connected? Will I belong? Will my instructors be responsive? As instructors, we aim to foster inquiry, curiosity, agency, and growth, but we can't lose sight of what can best support the opportunity for students to learn – care and belonging.

What Nicolas and I (Mandi) realized in our many conversations was that we needed to share our *aha!* moments with other instructors, and in turn learn from others. Our workshops and dialogues with colleagues helped us define the importance of care. We found additional grounding in an article that summarized the work of Nel Nodding. She says when teachers provide caring encounters, students are receptive and attentive in a special way and the relationship becomes reciprocal (Smith, 2020). In addition, an article from Carl Strikwerda claims that faculty members are essential partners of the retention of students towards graduation and academic success, which can be accomplished through fostering care (Strikwerda, 2019). The research validated exactly what we felt was missing from the online learning space.

Caring is relational. In teaching, it is a practice where teachers develop a relationship with students and develop concern for a student's overall well-being and performance. This concern creates a sense of belonging for students in the online learning space and can be created through verbal and nonverbal expressions of care, knowing student's names, making an effort to get to know more about the student, being knowledgeable about student support resources, and addressing student concerns.

Creating a space focused on the pedagogy of care requires focused effort by the instructor to use class time or design experiences to develop relationships and care. When utilized, the learning space can be a safe one focused on empathy and collaboration. The design and the instruction will focus on building relationships, activities and assessments will be designed with meaning, and there will be a connection of the content to student's lives. This sense of belonging for students based in the pedagogy of care will lead to more successful academic outcomes.

Centering Connection in Design

Another major theme that arose was the need to go beyond simple exchanges or interactions and design for human connection. At the heart of any interaction is the learners or individuals. Humans are unique in every way and bring their own unique emotions, experiences, personalities, and understandings to the connection.

While Moore's law of interaction types does guide us to build in space and tasks for learner interactions (Moore, 1989), the intention of interacting alone does not capture the specific interaction experience that develops connection and honors the individual. Simply positing the conditions for interaction will not foster meaningful interactions or connection. For the learner to feel connected to the learning, peers, and instructor, those interactions must be designed intentionally to connect by honoring the individual's uniqueness and providing flexibility for learners.

Peer-to-peer learning is a powerful learning experience and tool for instructors to use in the online learning classroom, but without the consideration or designing for affective learner aspects or social-emotional learning (SEL), then we aren't truly designing a holistic interaction or exchange between learners. What student achievement that we are left with is driven solely by intrinsic or extrinsic motivations brought by the learner. We can see poorly designed peer-to-peer learning play out in Massive Open Online Courses (MOOCs) where attrition rates are not ideal and success is driven by motivated learners and not by deep and lasting connection with others.

The following stories that we share highlight these themes and contextualize them across the many educational settings that we have worked. Each story will provide a description of the educational setting, student experience, instructor experience, observations of humanizing online learning, pedagogy of care examples, and applications of centering connection in design work. These narrative stories highlight some design practices and strategies and at the same time challenge many modern instructional design practices.

Being Seen, Heard, and Valued

Part of my (Mandi) K-12 teaching career was spent teaching science through problem-based learning in a middle school classroom at a STEM school. The school housed kindergarten through eighth grade, and the demographics were fairly diverse. It was a neighborhood school serving a population that was 72% Latinx, with a 79% free and reduced lunch rate, but it also included about 10% white affluent students from other areas of the district who had opted into the school for the programming.

Problem-based learning is rooted in cooperative learning groups, and as we all know, middle school is that awkward time where peer relationships can be sensitive and difficult. These cooperative learning groups were tasked with designing solutions to current world problems, organizing their thoughts into a presentation, and then delivering their solution presentation to experts in the field. The thought of middle school students of varying backgrounds being organized into groups and delivering a coherent presentation to a panel of experts was anxiety-inducing for both the students and me, their teacher.

After a few attempts at groups that fell apart and presentations that crashed and burned, I realized I needed a new approach and started on some research. The students in my classroom were from many different backgrounds with many different life experiences. They didn't understand each other. In order for student groups to work together to come up with successful solutions to a problem, the group needed to work cooperatively. Before a group could work cooperatively, students needed to feel a sense of trust and belonging within their group. Before students could feel a part of a group, they needed a lot of preparation with social-emotional learning.

At first, students were apprehensive about an initial strategy I used from restorative practices, called connection circles. We used it as a weekly check-in and check-out. It was a way for every individual student in my class to be seen, heard, and valued as we all stood in a circle.

I would give students a prompt and each student was given space and time to respond. The only person who could be speaking in the circle was the one holding the talking piece. The talking piece was a highly prized possession of the facilitator of the circle and not just a random object from the classroom. The point of the object being a personal prized possession is so the facilitator could model the vulnerability in sharing something personal with the students. After I described why the talking piece was so sacred, I gave students a prompt. We started out with silly but telling prompts such as naming a superpower that students would love to have. Spoiler alert, it was invisibility, because middle schoolers. Students did have an option to pass when it came to their turn, but it was not an exemption from the activity. Once we had gone around the circle, we would come back to those who had passed for their time to share.

Over time, the prompts transitioned to more personal items such as hopes, fears, and goals. I noticed students started approaching each other more about what had been said in the circle and started conversations about things they had heard. They started connecting with each other more on their own. It wasn't long before students began to ask me when we were doing another connection circle because they had something to share or wanted to hear what others would say. I used other strategies as well, such as using children's books and other team-building activities to teach empathy and growth mindset, but they all centered around building relationships and teams.

The environment became a safe space, a place to take risks, and to share new ideas without fear of humiliation because they had learned to listen and to have empathy towards each other. I taught them sentence starters to use with team members such as "I respectfully disagree because..." to help walk them through how to have appropriate disagreements. We changed the classroom environment to have care in mind first, so that the learning could happen. At this point, no matter what group I assigned students to, there was always some connection they could make to another student. The groups functioned in a much healthier way. Creative brainstorming sessions for solutions were much more creative and productive because most students no longer feared sharing a crazy new idea. Design thinking activities produced many more iterations because students had developed a growth mindset about the process. Because students had developed positive relationships with each other, they were more willing to give and accept critiques of their work. It was okay for the work to not be perfect the first time presented. Presentations were delivered in a more confident tone because they had the practice of presenting in front of their peers in the connection circle. Utilizing these social emotional learning practices led to a deeper academic experience for students.

The process of building social emotional capacity in middle school students was a huge paradigm shift for me and it became a very big part of my classroom. It was also a huge paradigm shift for me in how I approached my teaching practices and planning in general. Students craved connections with me and each other in the classroom. For the majority of students I saw the impact it made with their engagement in learning, their confidence, collaboration, and their social skills. I became very intentional about planning for the connection between students first in all my lessons and the content followed.

The extra time spent developing the social emotional aspect of middle school students was incredibly humanizing for them. It created a safe space for them to be vulnerable enough to take risks, share ideas, and work with others who were very different from themselves. Doing the work with social emotional learning first led to greater engagement with the material and learning of the content. It was almost as if I could get them to learn more of the content by focusing on it less.

Creating care and a sense of belonging in the classroom is necessary for students to be successful in the academic environment. They need to know that the classroom space is safe to take risks and develop a growth mindset. My STEM teaching experience taught me that it is necessary to explicitly plan with student connections in mind both in the design of the curriculum and the delivery of the material. If we create spaces for students to develop empathy to be good collaborators, they will grow up to be adults who have empathy and are good collaborators.

Bridging Success with Care NOT Grit

During my first graduate certificate program, I (Nicolas) taught in a dropout recovery high school serving a large metro region. Within this school resided 3 unique secondary high school models logistically designed to serve more students. I taught Career and Technical Education (CTE), mathematics, and science to 9th-12th graders in the hybrid model. The hybrid academy combined online learning classes that were self-paced with a half-day in-person portion that was required for most students.

These programs by description and design were to provide flexibility while still fostering persistence and "grit." A major emphasis was on the completion of coursework and consistent effort on the student's behalf. At the same point in time, the wave of "grit" and "persistence" was making its way through teaching conferences and teaching professional development. In this context these were the toughest kids I had met and it seemed redundant or frankly rude to attempt to frame their continued effort in the development of grit.

What worked better was developing relationships, connecting, and caring. This looked like consistent communication and strong relationships with parental figures in the students' lives. Being authentic and true to oneself drew out the student's personality and quickly built trust which led to better in-person attendance and my ability to keep students on task in the virtual setting. Trust fostered through care and authentic connection, drove student success so my task was to design for opportunities to develop trust and connection.

The classroom was a mixed grade-level experience with students taking short courses and gaining help with their online self-paced courses during the same span of time in the morning. This innovative model permitted students with limited transportation or access to successfully complete high school and earn a high school diploma.

For students, they could come to class and have connections with peers and teachers or borrow laptops and generally engage with each other. They were also required to attend a certain number of days a week to remain in the class which was often the most difficult aspect of the program. Many students had life situations that made this difficult or were not generally interested in school any longer.

Although this model brought flexibility and also a choice for how to attend class for students, some students needed time and space to engage while others just needed constant direction and support. Different student learning needs began to break this model down. This lab-based, in-person class time coupled with online, self-paced courses separated the learning from the human connection which caused student frustration and disinterest. It was often a struggle to make online learning and course modules relevant and not just webpages to be clicked through.

Outside of the 4-week short courses, the curriculum and course materials were provided on learning management systems. These courses were approved by the district and aligned to common core standards. This meant that when we were not direct-instructing or guiding our short courses then we were coaching, elaborating, and supporting our learners through their online courses.

In the support and relationship-building role, we spent time connecting with students and getting to know their life scenarios and goals. This often included finding out what they did for fun and why they were expelled from their original high school. These student motivations and moments helped to build a connection and a sense of care.

For the short courses, we developed very rich and engaging, peer-to-peer lesson plans and collaborations across Google docs and old school chalkboards. Students experienced problem-based learning labs related to topics like climate change and architecture and worked on them together. Journaling about their experiences and self-evaluating afterward helped them to connect the materials and experience even more.

By design, this model humanized the educational experience and provided flexibility through a hybrid model with lots of opportunities for engagement between instructor and peers. This needs-based approach takes out the structured and limited schedule of normal K-12 and provides an opportunity for a more meaningful experience.

When students trusted us enough to answer their cell phones or come to class FOR CLASS, that was success, that was perseverance. To build a relationship we had to build a culture of caring. This showed up in consistency of communication, check-ins, partnerships with their families where we highlighted wins just as much as issues, and making the learning relevant to their interests and needs.

With the application of pre-fabricated and pre-designed online courses covering common core curriculum, designing for connection showed up in two ways; the in-person time, and the connection of coursework to life and the workforce.

The in-person time was the space where we developed relationships and a culture of care as students worked together or studied together and built community in the classroom. Balancing a lab-like class with a breakout short course occurring in the same space was not always easy but it provided so much opportunity for connection. The students that would show up to study would see the fun and engaging activities of the short course and then sign up to join in on the short courses offered next block. The care and connection of the in-person portion humanized the learning and with that came lots of student achievement and high school diplomas.

Face-2-Face with the Pandemic

The Spring 2020 semester started like all of the past semesters with in-person classes and new students. My (Nicolas) course served as an introduction to the Teaching English to Speakers of Other Languages (TESOL) program so lots of new information for some and a rehashing of old for others. The student's backgrounds differed significantly with some paraeducators, 18-20 year old community college students, k-12 online tutors, and in-service teachers spanning k-12. Students enroll in this program because of our in-person courses although the program had been planning to adopt a hyflex course delivery model prior to the pandemic to expand enrollment opportunities to distance learners.

Since this course focused on the science of language and its intersection with teaching, I had designed a flipped course experience using a learning management system to deliver the readings, lecture videos, and low-stakes assessment prior to the in-class/in-person. The in-class time was spent on group work, discussion, think-pair-share, teach outs, and connecting the students to each other, the course topics, and instructor.

All that to say, a major value proposition of the program and student experience was the practice and engagement that would simulate and prepare students to teach. Many students would drive up to 100 miles to attend our classes on Saturdays. The program scheduled our classes on Saturdays to accommodate teacher schedules and provide a one-day school week for those driving to class.

The first eight weeks of Spring semester went as planned but during the second week of March, the global climate and public health orders were shifting. In week 9, the course went fully online and an extended spring break was given to accommodate for the in-person to online shift — remote learning had finally begun.

Suddenly students were required to have internet access and a computer throughout the week. This was a stark shift and not a requirement prior to the pandemic. Student relationships, communication, and collaborations also shifted into the online course discussion boards, IM, and emails. The collaboration and relationships that they had forged in the classroom were suddenly changed and changed in a fashion that was never desired.

When the decision came down to switch everyone online, I wasn't honestly too stressed or concerned from the design perspective. I had many things going for me. I am an instructional designer and faculty developer and my course was already flipped, so my LMS course was already developed in a consistent structure.

I began to plan discussions and assignments to simulate the types of practical assignments and creative opportunities that my active learning, flipped class normally provided. I began adding discussion forums with the fairly common interactional structure of, first post by Wednesday and replies by Sunday. My focus was on making the initial prompts as rigorous and appropriate as possible and made the assumption that the required replies would facilitate a community of learning and connection — I was wrong. To help facilitate a smooth transition, I shared the following course expectations with my students:

> Hello Class,
>
> Today begins the final 5 weeks of the course and as you all are hopefully aware no more in-person class sessions. To make this switch from a hybrid course to a fully online course, I have redesigned the final 5 weeks to be consistent and very similar in structure. This will help you succeed!
>
> **Course Tasks and Expectations**
>
> For the remainder of the course, each week will consist of:
>
> - Discussion where you will be required to reply by Wednesday at midnight and then respond to two of your classmates by Sunday at midnight.
> - Assignment where you will be asked to find an ESL activity or instructional material like video that works on or is focused on the weekly topic and submit a quick analysis.
> - We have two more major projects due in week 12 & week 15 – in those weeks there will not be an assignment.
>
> This is what will be required of you to pass the course moving forward. I have removed quizzes and any additional assignments. What is listed above and in the weekly modules moving forward will be what is expected of you.
>
> To help you plan your weeks to succeed with this new online schedule, below is a helpful schedule to keep you on task. I would suggest adopting it!

Key to succeeding in this course for the last 5 weeks

- OVER-Communicate with me! Like way too much. We can't communicate enough!
- Complete the 5 weeks of content and submit any assignments that you may have missed form the first 10 weeks.
- Contact me with any questions or concerns

~Nicolas

Five weeks later, after countless IM and emails to students and discussions with colleagues on topics of "how much communication is too much?" I realized that through the shift, I had lost some of my students and my class community had mostly disappeared. Many colleagues in higher education were feeling this loss of engagement and experience with students and turned to blaming concepts like remote learning vs. online learning, but for me it was a loss of connection. The cold digital version of "community" that this LMS supplied did not support a rich and engaging space of interaction that made individuals feel human and connected innately. I was unable to transition a sense of care and connect with my students.

Fostering care in an online learning environment requires some significant focused work and facilitation. We can not create a sense of care or actual connections when leveraging talking head videos or using first names in our discussions alone. We must design for care and center efforts to connect with each and every student.

Prior to the shift to online learning, students would enthusiastically attend class on Saturday knowing that they would be presenting, sharing, collaborating, cooperating, and teaching. They felt connected across Moore's interaction types and they felt heard and cared for through my actions and follow up as the guide in the class.

When we moved to online learning, the design norms, functionality, and connection of discussion boards and the all too common, "post your initial response by X-day and respond X times by X-day," failed to produce a sense of connection. My major takeaway was to never look at online student-student interaction the same way and to design group activities with meaning and connection centered. When designing student-student interaction we can't count on technology to bridge the connection, we must do that in design and in facilitation.

Through this experience, I learned that sound instructional design focused on outcomes and deliverables of student performance can not replicate conditions for learning alone. We must design for care and connection and ultimately humanize the digital learning space. We need to create opportunities for individual expression, sharing, and choice whether the course be online, face2face, asynchronous, or hyflex.

My design work to humanize the digital learning space took on new technologies and included modifying existing practices. My students needed opportunities to demonstrate their understanding but do it their way. The simplest modifications were shifting discussions to video discussions via Flipgrid and Padlet, where students could bring their personalities and interests to the digital space. In addition, I applied principles of differentiated learning to my assessments. My students needed choice in their ways of demonstrating learning outcomes and making the learning relevant and real. This looked like rubrics that provided choices for videos, PowerPoints, essays, research papers, recorded presentations, or live presentations. These design changes brought the student into the digital classroom much further than my previous practices.

Relationships First, Content Second

In addition to my role as a faculty developer and instructional designer on our university campus, I (Mandi) am also an instructor in a Bachelor of Arts Completion Program; a program designed for students to finish a previously started four year degree. The students in the program have a wide variety of backgrounds and varying success in an academic settings. When I agreed to teach for the program, I recalled my own experience of completing a certificate program beyond my masters degree in an online environment. I felt entirely disconnected from the school and most of my classmates even though we spent 3 semesters together in class. I was determined to provide my students with a different experience.

When I received my first teaching assignment, I read through all the material in the course and realized that everything students needed to learn the material was right there for them in the course. If an instructor wasn't present, could a student still learn? I began to wonder, what did an instructor add to the online learning space? When I asked an instructional designer on my team about it, they listed the ways communication from the instructor fit into the design. The conversation gave me pause because it seemed very backwards. Shouldn't the design of an online course be centered in fostering communication and connection between the instructor, students, and content?

I immediately set out to plan for ways I could bring the humanizing aspect into the online setting. I wanted to get to know my students better, find ways for them to interact with each other in the spaces provided, and show them I cared about how they were doing in class. In my introductory discussion, I used an online tech tool called Padlet to create a self profile to share with students. It was created with pictures of me and family, links to interesting sites and hobbies, and who I am as a learner. Students also created learner profiles, complete with pictures of pets and other creative contributions. In the discussion boards, I directed students into small conversations with each other by pointing out similar thoughts between students and asking probing questions for them to answer, I used names and specific feedback in the gradebook, and I even held and recorded optional synchronous sessions on Zoom to answer questions about assignments or clarify the content. It still did not seem like enough because I did not feel the same connection to students in this environment that I had in a classroom environment.

Student evaluation feedback was overwhelmingly positive. Students commented that they were excited to be in the discussion boards because the conversation was so engaging. They felt they really learned a lot from the feedback and the material was applicable to everyday life. Students mentioned the fact that the recorded videos of the live sessions showed I cared about them and their unpredictable schedules because they could view them whenever they wanted.

Even though the majority of the feedback was positive, I still had a student mention that they didn't feel enough of my personal attention was given to them in the class. I know we like to dismiss these students as one of "those" students, and there's one in every class, but that is a concern to me. Our university offers hundreds of online courses each quarter, and if there is one student in every class that doesn't feel connected, that's thousands of students every year missing out on a sense of belonging.

The beautifully designed, but fairly rigid structure of our courses does not create enough space for human connection. We need to design more time for solution-centered conversation around humanizing the space. Students are one of the biggest stakeholders and should be brought into the process. There have to be more creative options available than to post an initial response by Wednesday and respond to two others by Sunday.

Should design be about predictability and routine which in theory reduces student strain or should the course and intention be more about how we connect to students? By way of less focus on routine and more focus on students? In a learner-centered, care-centered world, the focus should be on the student and their own individual learning process.

Creating the space for students to share themselves with me and others in the class is a way to bring humanity into the online learning space. It inspires empathy and care in a seemingly cold technological setting, and has made me an advocate for designing and delivering courses with the pedagogy of care in mind. It's the first thing I introduce to the faculty I onboard and should be in the forefront of every course design.

Ensuring the humanization of online technological spaces means we need to design courses with opportunities for human connection in the forefront of the design. The delivery of the content also needs to be intentionally planned with opportunities for engagement between the instructor, students, and the content. The leap from K-12 into higher education has taught me that students in an academic setting, no matter their age, have the same needs from their teachers; to feel seen, heard, and valued. It begins with thoughtful design centered in care.

Connecting & Caring in Summary

As designers and learners ourselves, we are still working to answer our initial question, "What does the experience of humanized online learning feel like for instructors and students?" From our reflections and observations, we strongly believe that we must center care and connection to humanize online learning. Throughout these stories we identified a few critical anecdotes to designing online learning informed with care and connection that fostered student achievement. We hope that these anecdotes will help you to design online learning that is more caring and connected in nature.

If online design does not permit for care or connection then the design must change. We must humanize it. To humanize online learning is to teach and design with compassion. Caring is a reciprocal action which extends beyond the instructor and student and seeps into all student relationships. We believe that this approach to teaching and learning is pivotal to humanizing learning and we could say that learning guided by care is central to the cultivation of care in society.

References

Moore, M.G. (1989). Editorial: Three types of interaction. *The American Journal of Distance Education*, 3(2), 1-7.

Smith, M. K. (2004, 2020). *Nel Noddings, the ethics of care and education'*, The encyclopedia of pedagogy and informal education. [https://infed.org/mobi/nel-noddings-the-ethics-of-care-and-education/

Strikwerda, C. J. (2019, September 4). Faculty members are the key to solving the retention challenge. *Inside Higher Ed*. https://www.insidehighered.com/views/2019/09/04/faculty-must-play-bigger-role-student-retention-and-success-opinion

Developing Critical Student Autonomy in Blended Learning
Perspectives from a teacher and student at African Leadership University, Rwanda

Laurel Staab and Martin Wairimu

How might we design courses that create inclusive communities where students are able to engage in critical discourse and build their own learning experiences? This was the question at the heart of the design of the first Advanced Education Seminar (AES) at African Leadership University (ALU) in January of 2020. This piece is co-written by Laurel Staab, the designer and teacher of the course, and Martin Wairimu, a student who was enrolled in the course who helped with its conception and design. We provide an example of how Critical Digital Pedagogical theories can be applied in practice in the design and delivery of a course. We start with an explanation of the context and theories that informed the design, then present an overview of the course design itself. Next, Martin provides an analysis on the student experience of the course. Martin conducted informal interviews of his peers regarding the course a year after the start date of the course, and has integrated those perspectives into his own critical analysis and reflection. Then Laurel provides her reflection on the experience of the course from the perspective of a teacher. Finally, we conclude together with our shared observations and recommendations. We hope that this discussion provides some insights to other educators and students who are interested in designing learning that encourages high levels of student autonomy in blended learning environments.

Background

ALU opened its second campus in Kigali, Rwanda in September of 2017 with an undergraduate class of 270 students from over 30 different African nations. By January of 2020, two more classes had joined the school. The students enrolled in the AES were in their final year, and had chosen to study a project-based degree called Global Challenges (Staab, 2020). The course was an advanced elective that thirty students within the degree selected because of their interest in education, and it was the first time that the course was delivered. The majority of students enrolled in the course had taken an intermediate Education Seminar the previous year, where we had studied critical pedagogy and constructionism.

We intentionally designed the course to be self-directed, project-based, peer-driven, and to leverage a blended learning model that incorporated both in-person and online components. All courses at ALU at the time used a blended method of delivery with elements of both in-person and online learning. Students have their own laptops and access resources and submit assignments via an online Learning Management System (LMS). From its founding, ALU viewed the role of the professor as a "facilitator" of the students' learning, guiding and supporting students towards their own goals rather than setting those goals for students (Faraj, 2019). However, we struggled to practically implement models where students have high levels of autonomy over what they learn and how they learn it. The AES was my attempt to push further towards what I will call "critical student autonomy," or a degree of autonomy in which students have high levels of control over what they are learning, how they learn it, and how they demonstrate what they have learned.

The first challenge is that critical student autonomy requires a departure from how most education is designed and delivered. Most often a combination of the teacher, the school, and the state choose what students should learn and how they are assessed (Giroux et al., 1988). There are, of course, reasons for this. For one, some might question whether students have the proper expertise to choose what they should learn. Teachers are supposed to know more than their students, and formal education has often assumed a transfer of knowledge from those who know more to those who know less. While ALU set out to resist this model, our institution has struggled with the concern that students might not reach the level of depth that is needed for critical engagement without a teacher-centric approach. Furthermore, from a national perspective governments and regulatory bodies often decide what knowledge and skills their citizens should possess in order to contribute to society, and learning institutions must comply with these mandates.

But for as long as there has been this standardized model of education and learning, there have been alternate theories and practices that challenge these conventions. Paulo Freire presents an important critique of the "banking system of education:" the idea that students are empty repositories that teachers can "deposit" knowledge into the brains of their students (Freire, 1985). When you view traditional education through this perspective, it indeed becomes problematic, undemocratic, and even dehumanizing to view students as passive recipients of knowledge with no autonomy. Critical pedagogy scholars also point to the connection between the banking system of education and the broader project of colonization, itself a dehumanizing project that has had a lasting impact on education systems worldwide (Giroux, 1985; Katundu, 2019; Nyamnjoh, 2012). Therefore, proposing an alternate way for students to participate in education where they have autonomy over their learning experience is a part of the broader project of decolonization of education, something that I was very concerned with as a white US citizen teaching students from across the continent of Africa.

Another important perspective that leads us to question traditional education design and delivery comes from the many scholars and practitioners of education who have made it their life's work to study how people actually learn. Like the critical pedagogues who question whether knowledge should be deposited into students' heads, learning scientists and theorists question whether it is the most effective way for people to learn (Immordino-Yang, 2011; Schwartz et al., 2016). At ALU, our learning model and broader education philosophy attempts to align with learning theorists who have questioned rote learning and the banking system of education. We believe that students should have opportunities to learn by doing, whether through experiential and project-based learning or more self-directed opportunities to build and construct knowledge through action. For myself as an educator, I drew heavily on the work of Seymour Papert and his theory of constructionism when conceptualizing how to create learning experiences, where students have the opportunity to design and create projects of their own choosing related to a personal mission that they have chosen for themselves. Whether it is building a tangible engineering project or designing a curriculum, projects allow students to learn through building, applying theoretical knowledge that they have studied towards the practical creation of a product, and linking their theoretical understanding to the concrete (Papert, 1980; Resnick, 1998).

Designing for this level of student autonomy is also often at tension with systems that demand standardization. However, it predates more recent learning theories and models for schooling and aligns with pre-colonial education models in which most education happened through apprenticeships, some formal where the learner practiced a given trade and worked with a more experienced person to learn the trade, and some less formal where children learned skills with family members and those around them (Ezeanya-Esiobu, 2019). However, while there can certainly be room for creativity and autonomy in apprenticeship and situated learning models, these models can at times be more controlling depending on the needs or requirements of the apprenticeship or work environment, while self-directed learning by definition requires high levels of student autonomy and ownership, both on the part of the student and by the intentional design of the educator (Hiemstra & Brockett, 2012; Lave & Wenger, 1991).

Design of the Course

I had previously implemented some aspects of a self-directed model for teaching and learning in different courses and programs that I designed, but was excited to push the boundaries of student autonomy with this particular group of students, many of whom I had previously taught. I also was excited for the opportunity to co-create the course with student input. In the semester prior to the AES, Martin was working as an intern for my department, and he interviewed several of his classmates about their previous experiences in an education course that I had co-taught, as well as what they wanted to learn and do in the AES. When we analyzed the responses together, we found that students wanted opportunities for practical learning, to learn how to design curriculum, teach, and start their own education programs. We discussed this feedback, and given the many ambitions, we decided that students should have as much autonomy to learn and do what they wanted as possible, but with the intent to provide sufficient structure and scaffolds to help them stay on track in terms of motivation and accountability, one of the greatest challenges of self-direction (De Bruin, 2007). Therefore, there were three critical components to the course that I designed.

The most important component was the projects that students chose to work on. At the beginning of the term, students planned the projects they would complete, chose the learning outcomes that they would master from a menu I provided, and scoped out work plans for how they would complete their work. When conceptualizing what their projects might be, I thought that students might most commonly design curricula for different programs and learning experiences, or might also design broader education projects, theories of change, and workshops, as well as individual lessons. In order to help students conceptualize and concretize what they might choose to create, I provided them with three pathways within which they could design their deliverables, with the option that they could also self-design another pathway that did not fall into my defined categories. Students were meant to meet with me at least three times during the semester so that we could discuss their choice of deliverables and progress.

The next component was the Weekly Learning Journal Reflections. Students were required to write an entry each week recording what they had read, what they had discussed with their peer group, what work they had completed towards their deliverables, and most importantly, what they had learned that week. This was meant to build their critical reflection and metacognitive abilities, and gave me a good way to quickly gauge who was doing what and what was being learned. I used the completion of the weekly journal entry as the measure of attendance in the course, rather than attendance at weekly sessions, to emphasize the importance of self-reflection and personal learning as the key purpose of the class.

The final key component of the course was the peer groups. ALU relies on peer collaboration starting in the first year, but I found that by the time I teach students in their second and third year, many are exhausted by group work and projects. Instead of focusing on group projects, I emphasized building a peer community called Communities of Practice (COPs) where students supported each other and held each other accountable for their individual projects. Students self-selected their COPs in person during the first week of the course based on similar interests. They spent time early on completing team building exercises and drafting a group manifesto that established their group expectations and rhythms for meeting together. Groups were encouraged to meet weekly and discuss readings that they had self-selected, as well as discuss their individual progress on their projects.

I de-emphasized the importance of traditional in-person classes for this course, and made all in-person classes optional, but we did meet for synchronous sessions in-person once a week and transitioned to online synchronous sessions when we switched to fully online delivery over the course of the pandemic. During the first week of the course, I surveyed students to see which topics they were most interested in learning about in the synchronous sessions. While my idea was to have sessions "on demand" when students needed to learn a concept or skill based upon their project work, the on demand resources were not as adaptive to student needs as I had initially conceived because student interests and needs diverged greatly, and I struggled to break from the standard class session format that I am used to as a teacher and to have diverse enough offerings to meet all student needs. This was definitely an area in which the course design could be improved, though over time a library of sessions and workshops could be built.

Student Analysis

When Laurel revealed the course design at the beginning of the term I could see mixed reactions from my peers. The most common reaction was curiosity. Questions ranged from, "What is this term going to look like?" to "Is this for real?" One riveting component was the fact that attendance at the "in person workshops" was optional. It felt unbelievable, as classroom attendance at ALU was mandatory. It felt unusual; a mixture of freedom with uncertainty. I was a little unsure if I was ready for this type of learning. I could feel the same reactions from my peers in the classroom. The following were the thought-provoking aspects of self-directed learning that Laurel announced in class that day.

1. Mission Meets: Students were to meet Laurel at least three times in the term to share the progress of their mission project/product.

2. Optional class workshops: With no attendance record, attendance was the mandatory weekly reflections that was to be written in the learning journal, where students shared what they learned that week.

3. Structured Choice: Students selected learning outcomes that they wanted to be evaluated on based on the products they would create.

4. Communities of Practice (COPs): We also chose peer groups, and diversity was encouraged. COPs were a form of peer learning where a group of students who shared interest in respective fields within education would come and exchange what they had learned during the week.

5. Workshops: We had two in-person workshops in a week. On Monday evenings there were facilitated sessions based on the students interest. On Wednesday evenings, the COPs would meet in the classroom space and the facilitator held mission meets with individual students at the same time.

6. Work Plan: We were to submit a work plan for the class with the timeline of when to submit our individual products and the learning outcomes that we were observing.

Choice and autonomy were the words that summarized the AES's self-directed learning experience. This included the freedom to choose what I wanted to learn, decide how I wanted to learn it, and determine how I was going to be evaluated. I had complete ownership of the learning experience in that class. One of my classmates put it best when they stated, "I feel like my entire ALU life it's only during the [AES] that I really had what we called self-directed. I understood that education is not just having a curriculum, or just having good content. There are a lot of operations in space that I need to use as keys." Below are more highlights and aspects of the course that made it an effective learning experience.

Ownership

Ownership in the self-directed learning that we experienced includes but is not limited to autonomy of learning experience, journal reflections, and COPs. Self-directed learning allowed us to choose the learning tracks we wanted to delve into and subsequently develop two products. The class structure allowed independence in attending optional sessions customized to what students desired to learn as well. However, a fellow colleague remarked on a challenge with this model, saying that "we were learning a particular topic about the course but not everyone was paying attention to everything. Everyone was focusing on what they were doing for their summative. So I'll recommend that everyone does a formative for every topic whether it's for the curriculum design, or facilitation, or the workshops. In that way everyone is learning everything." For many of us this worked because it allowed us to go deeper into areas of our interest. In this view a colleague remarked that "I feel like I was able to learn, like a lot of things that I wanted, and then the whole program was designed in a way that it guides me, like, I may have a passion for these things that may be all over the place. But the [AES] would just give me like, you know, help me put some order." Self-directed learning in the AES was designed in a way that it gave us a choice of what we wanted to learn as students and how we were going to learn it. The facilitator also gave us a liberty of selecting the deadline for submitting our respective products. The facilitator gave us a range of dates where we could choose to submit our work. In the end we were to provide products as evidence of what we had learned throughout the course.

Communities of Practice

COPs were an extension of ownership in self-directed learning. We had the opportunity to discuss books, resources, and certain topics in the field of education with minimal supervision. On Wednesdays we'd come together to reflect and share what we had learned from our individual readings and resources over the course of the week. It was an amazing experience learning different perspectives on a subject from my fellow peers. For me, it was an exciting feeling just experiencing a sense of community from my peers who have the same enthusiasm for education as I do. A colleague put it best when they said, "I went above and beyond to utilize the capacity of my [COP] by tapping into the knowledge of my community members. It was actually a community that we wanted to continue with, to, to continue staying in touch with one another after the end of the course. So, we actually continued working together and sharing opportunities and ideas together. So that sense of togetherness is what kept us going. And the teamwork was amazing. You know, like, how I interacted with my peers, it was as if, all of them were, like, my contact was on your speed dial." The COP effectiveness in the whole education AES learning experience can also be summarized in this quote: "I am a person who does not enjoy peer work at all. However, this one is unfilled. This is more like a book club. So we got to the people we were with, and we got to align." The students engaged each other in critical discussions that they were passionate about and it was very productive.

The Learning Journal Experience

The facilitator required students to write weekly reflections and share what they learned either during their class sessions, individually when interacting with the resources shared, or during COP meetings. The facilitator shared a Google Doc with a reflection journal template where we made our weekly updates. Since attending in person sessions was optional, the weekly reflections were compulsory. The learning journal experience was unique; it forced me to think and actually made me take the course seriously. Moreover, for some students, it was a journey of self-discovery and introspection. One of the students remarked, "every time we got to reflect, you could find out maybe if you're not making sense, you see, like, the more you reflect, you can tell if you're making sense if you're not making sense. And you are able to internalize even things you'd not known." That sentiment was echoed by another student. "[It] was a very good experience of zooming everything, putting it there, and understanding why you do what you do. And it also made me reflect on my personal strength. And the reflection was able to make me understand what are the skills that I need to get... Because just to help me reflect, and I get to see where I was struggling, where I was stuck, and really understand, like, what do I need? Which type of skill do I need? For whom do I need to partner with?"

However, the learning journal experience was not as effective for everyone, as the following quote demonstrates: "It was more of something to check off for me because I was supposed to talk about what I learned that week, what we as a peer group discussed in our meeting, and also my takeaways from the week's topic. Just because this was a self-directed learning course, it made me put everything else first and put this course last because I had my own deadlines and I was doing everything my own way. It kind of lowered my effectiveness. I knew that I was supposed to submit my journal every Sunday so I was doing everything that night including the readings."

Some students found it very introspective, where you get to see your growth points and strengths while observing your learning journey. For some it became a tedious check of tasks that reduced the effectiveness of the learning process. I loved it since it forced me to go back to my notes and actually reflect on what I had covered that week.

Challenges

When the pandemic struck in March of 2020, it disrupted the course of things, and the AES wasn't left behind. We had to transition to fully online learning. The transition wasn't easy and this marked the biggest challenge in the course. First, things changed very quickly. One day we were in school and the next week some of us were traveling to our home countries. Secondly, we were so used to seeing each other and having those conversations in class, but now we had to have them online. The change was startling and unsettling to put it best. The internet issues, adjusting to the home environment, and learning became so difficult.

A fellow student explained, "It was very frustrating. As much as we talk about self driving, I just saw the importance of having people around me and having people in class…I wasn't prepared for this…It was a mess. I wasn't ready, the in-person discussion seeing physical people around me was good and online was very hard…the amount of time you have to invest in an online relationship is two or three times more than a physical one." The full transition to online learning was very difficult and it was sudden so we felt very unprepared. In spite of the support from the class facilitator, it took time to fully transition to online learning. These challenges also translated to some of us not being able to put down our weekly journal reflections. The facilitator responded to this challenge by providing some more time and reducing the weeks that we were required to have journal entries. She reduced them to ten entries during the fifteen week semester.

Furthermore, during the course some students didn't have enough time with the class facilitator, the one-on-one mentor meets. Some students felt like they could have had a greater learning experience if the mission meets were mandatory as this would have pushed them to do more and get more support in their learning experience. Mandatory mission meets could have allowed for more individualized support in student's learning progress. I'd also have loved more mentor meets, as this would have also pushed me to be more critical. A challenge I experienced was the feedback on my side took a bit of time. I had done my products and had observed my individual work plan well, which was quite ambitious as I remember it. Feedback came later during the end of term and if it had come a little earlier I would have had time to work on it and thus improve my overall learning experience. This is however, something that was limited to me, as other students did not share this remark.

Some students also felt that learning journal reflection guiding questions were limiting the scope of their learning experience. The learning journal had guiding questions that students could use to write their weekly learning reflections. "The guiding questions were limited, it should be a free write of what you learned." In addition to this challenge, when it came to choosing the learning tracks some students felt that this also limited their learning experience. They recommended more regulation, that every student should study what everyone else is studying in the class to increase knowledge of education in other subjects.

Critical Pedagogy

While critical pedagogy was implemented in practice more than it was in theory, there were several ways that it shaped our experience of the course. First, there was the way the classroom was arranged during sessions. We arranged tables in a circle so that we could all see each other, including Laurel. This reinforced the role of the teacher in class as a guide and a learning companion and challenged the hierarchy between teacher and student. There were also multiple course resources that discussed critical pedagogy. One of them talked about indigenous knowledge systems in Africa, where we critiqued the curriculums we had studied in most schools across Africa. The practice of discussing these resources among the COPs brought about an inquiry that we needed. I remember investigating the identity of authors questioning their background and motivation for writing. Ultimately, we questioned whether the education systems in Africa were meant to empower or reinforce colonial dominance and influence, the latter being the conclusion.

Additionally, in our COPs we were required to be active which involved drafting a document with norms and ways of working as well as roles for every member of the group. I remember the teacher emphasizing that no person should take one role all the time. During our first month, I was the scribe, in our second month, I was the time keeper and in our third month I was the moderator of our discussion. This way everyone took responsibility and we shared leadership. Finally, we had a session on critical thinking and critical pedagogy. One particular question that Laurel posed in the group was, "Do you think a teacher should give their opinions in class?" I remember many peers responding to this question by agreeing that teachers can share their opinions; however, they also expressed that the teacher should be careful to either not dominate the discussion and allow for views that are different from the teacher's. It was a great discussion where some peers expressed concern in cases when teachers in class shared information that is not factual and questioned who could hold them accountable in that space. We also discussed curricula and questioned whose voices were represented. Does it reinforce oppression of the minority and marginalized, or does it empower? It is my belief that critical pedagogy was understood as the practice of removing a teacher's dominance in class and any forms of hierarchy in a classroom, questioning the contents of a curriculum and reinforcing student's voices.

Student Reflection

Overall, this course was a great experience for students as they were able to create their own viable products solving different problems in education within their communities. Some students designed summer leadership camps and others worked on passion projects. Even though some students felt like they could have learned more, many students were satisfied with the execution of this course. One aspect that continually emerged was self awareness in learning. The reflection journals allowed the students to comprehend their level of proficiency or knowledge of respective topics in education. We also valued the fact that attendance was not mandatory and weekly journal reflections were introduced at a good level where students had mastered some sense of self discipline in studying. Moreover, a student also applauded the design and structure of the course. The student added that because of the design and structure of the course, the transition to online learning became smooth. Of course it could be improved to suit overall needs of the students and more feedback is needed. Based on the feedback, even with the challenges experienced, self-directed learning was successfully executed.

The whole course was different and I loved the fact that I explored my knowledge and skills in education. Most importantly, I challenged myself in working on my mission projects. An integral piece that most of my peers agreed with was the vast experience of our facilitator. I particularly enjoyed her questions when I attended mission meets and anytime that I had conversations with her regarding my mission project. She had a way of asking questions that would either give me clarity or lead me to seek more understanding and do more research on a particular topic. This is something that even my peers would experience. A colleague remarked that before I came to education class I felt like a lost sheep and after a conversation with Laurel, I found how education connected with my project. Such was the power of the conversations she would have with students. One other aspect was the fact that she shared many resources. I enjoyed the books and articles she shared. They were always nuanced and well thought out. Her approach to teaching was always student-centered. On a particular occasion she brought a guest speaker, our former head of college, to discuss education and leadership. It was meant to be a question and answer session. She took the liberty and drafted questions on our behalf, and as many commented, those questions were nuanced and covered the topic very well. In complimenting her, when she remarked that she took about five minutes to draft those questions, I only sighed in disbelief. What followed later was a promise that she would teach us a session on "Questioning and Eliciting." She covered it at that time when our hearts were terrified of the pandemic, though it was online, it was still one of the best sessions I have attended. To put it in context, I remember the icebreaker she used for that class. We used colors to express how we were feeling, followed by a question, "Why do educators ask questions?" The session climaxed with Bloom's Taxonomy pyramid chart showing the different levels of questions. Sitting in my living room in Nairobi, I couldn't help but feel nostalgia due to the fact that this experience could have been better if it was in person. Nevertheless, such has been the effect and impact of my facilitator's teaching experience. That session there marked the height of the course. It was the last session I attended with her, and it's still fresh in my mind a year later. I can confidently say that this course was successful not only because of the design but also because of the great support from the facilitator. Of course as a student, I was equally responsible for my learning experience, and I learned a lot in the process not only about my skills but also I became more self aware in the process, which was remarked by some of my peers.

Teacher Reflection

I wanted to write this piece because I observed from my role as a teacher that the way that I designed the course meant that it was relatively easy for me to transition from a blended to a fully online delivery model. This was in contrast to the other course that I was teaching at the same time that did not have the same levels of self-direction. The abrupt transition to fully online learning meant that I had to dedicate more time to administration of the transition to online learning for the broader university, and I had less time to support students. It was a stressful time, and I was not as prompt in my feedback to students in their journals as I intended and I did not follow up with students as much as I had intended to. This is something that I still feel guilty about. Yet despite my limitations and failures and the many challenges students faced, their final projects and reflections were incredible. When I began the design of the course, I had seen it as a test of a model that would practically implement many of the components of learning that I believe in: students learning through building their own projects, students reflecting about their process of learning, and students building supportive communities with their peers in which they could have deep discussions about topics that they were particularly interested in. Despite the many challenges that we faced during the semester, many of these components did work in practice the way I had hoped, and many students created impressive projects.

It is also insightful for me to be able to view and reflect upon the student research that Martin conducted. I feel proud to learn that some students not only gained value from their COPs but that they continued to sustain them beyond the course itself, as the creation of communities and lasting relationships is always a goal of my teaching practice. At the same time, there was valuable feedback about how students interpreted aspects of the design of the course that leave much room for improvement. The weekly reflections were a place where I as a teacher gained valuable insights into the thoughts of students, and I provided guided questions to support students in how to shape their responses. However, I could have emphasized that these questions were only meant to prompt writing if they were useful, and could be discarded if they felt restrictive or formulaic. Additionally, while I valued the regular cadence that the weekly reflections provided, having a required assignment that I graded and used for attendance runs the risk of becoming something that students complete simply because it is required for a grade and compliance with institutional requirements, and diminishes the goal of the critical reflecting that is the true intent. As these reflections were at the heart of the practice of building critical student autonomy, making these reflections ungraded and with a clear opportunity for students to respond in an unrestrictive format is the most valuable revision I could make to the course. Additionally, it is clear from student responses that there were varying levels of internalization of how theory had informed the course design. Most of the students in the AES had also taken the previous education course with me and had discussed critical pedagogy and constructionism, and we delved deeper into these topics during our synchronous sessions in the AES. However, we could have had more explicit discussions and critiques of the course itself during these sessions where we explored how well AES implemented principles of these theories to build critical student autonomy.

Shared Conclusions

When we started the AES at ALU in Kigali, Rwanda in January of 2020, we had no idea that a month and a half later a global pandemic would dramatically change how education would be delivered worldwide. But due to the nature of the design of the particular course, making that abrupt transition to fully online delivery worked surprisingly well. But what did we learn about learning design and critical digital pedagogy? There are several key components of the design that we think made the course what it was that we would like to highlight and discuss.

First, one thing that continually came up as the fundamental aspect of the course was that students had autonomy over their learning by choosing what they wanted to learn and how they would show what they learned. As a teacher, giving up that control can be a bit daunting, and for students, it can be more work and can sometimes feel intimidating. There is a tendency among both the teacher and the students to feel uncertain about high levels of student choice, as both may worry, "are students learning what they need to learn?" However, pushing through this uncertainty helped us both to challenge the idea of a standard curriculum that everyone "needs" to learn, and it can also feel liberating to witness and have that level of freedom. For example, one student who is passionate about the applications of virtual reality enrolled in an online course and began to learn how to design virtual reality gaming environments. This is something that the teacher didn't have the expertise to teach, but based on the course design he was able to clearly articulate what he had learned, the strengths and weaknesses of his work, and how he would continue to learn and apply his learning towards his final capstone for his degree.

Another key part of this course was the idea that students would learn through building and reflecting. This is meant to be an application of constructionism, and the heart of the course was students creating final projects and learning by creating these projects. The weekly reflections became the key place where students could articulate what they were learning through their projects, and the place where the teacher could see evidence that learning was happening. The repeated reflection meant that students built their reflective skills, and the quality of their final reflections was impressive. As a teacher I could see tremendous learning happening simply by asking students to tell me what they learned, and reading and asking students follow up questions or recommend relevant resources for them to engage with. As students we were able to document our learning and make sense of the work that we were doing.

Next, the technology that we used for the course was not particularly advanced. We used Google Docs to create the learning journals that the teacher and students could both view and edit, and ran synchronous sessions using Zoom and Google Slides. Our university was in the process of transitioning to a new Learning Management System (LMS), so we used Google Classroom for submissions and communication. Students used WhatsApp to communicate with their COPs and Google Meet when they chose to meet synchronously. The priority at the time was accessibility as many students were working with low bandwidth or hot spotting when wifi was poor or not available, and the technology was simply a tool for us to stay connected as we worked on our projects, met with our groups, and wrote our reflections.

The fact that the course began in-person meant that a lot of the initial community building as a class and in communities of practice happened in person up front before we shifted to online, which meant that we didn't have to engage in virtual community building which takes a significant investment. We have both seen and led fully online courses since then, so we know that this takes a great deal of work, but we do believe that it is not just possible but necessary to build strong communities virtually, though it will alway feel different from the in-person experience. Since these strong bonds were built, we were able to have deep conversations about topics like critical pedagogy and creating safe and brave spaces in education during synchronous workshops. The COPs were also a site for learning through community, though some were stronger than others. Students were able to get to know each other at a closer level, and learn about each others' passions that underlie the projects that they chose, such as why a certain student was motivated to start a community school or why someone else was passionate about early childhood pedagogy. The COPs had shared goals and helped to keep each other accountable in achieving their individual goals, which can be a challenge in online learning.

Students also had deep discussions in their COPs about articles and resources that they selected. One group that led a tutoring program that had been founded to help students read in English explored mother tongue instruction and the colonial ideologies that led them to prioritize English. The fact that the peer groups could select their own readings to discuss weekly meant that there was also autonomy to have deep discussions on different topics that the teacher had not necessarily selected or envisioned.

When we center the design and delivery of our courses around the assumption that our students are the best ones to choose what they will learn and how they will learn it, we are affirming their humanity. Whether this happens in an in-person, blended, or online setting is significant, but it doesn't change this fundamental assumption. Building relationships between the teacher and the students and among students themselves is not only possible online, but necessary for the type of learning experience that we desire. While many platforms and LMSs have been built to try to streamline learning and assessment, we found that in this course the simplicity of a shared Google Doc provided both a space for students to share what they had learned and a way for the teacher to quickly have visibility into what students were learning, while meeting one on one and in groups created spaces for critical questioning, dialogue, and community.

References

De Bruin, K. (2007). The relationship between personality traits and self-directed learning readiness in higher education students. *South African Journal of Higher Education*, Vol. 21, 2.

Ezeanya-Esiobu, C. (2019). *Indigenous Knowledge and Education in Africa*. Springer Nature.

Faraj, G. (2019, April 23). *Moonshot Thinking for Global Challenges*. Medium.

Freire, P. (1985). Reading the World and Reading the Word: An Interview with

Paulo Freire. *Language Arts*, 62(1,), 15–21.

Giroux, H. A. (1985). Intellectual Labor and Pedagogical Work: Rethinking the Role of Teacher as Intellectual. *Phenomenology + Pedagogy*, 20–32.

Giroux, H. A., Freire, P., & McLaren, P. (1988). *Teachers as Intellectuals: Toward a Critical Pedagogy of Learning*. Greenwood Publishing Group.

Hiemstra, Roger, & Brockett, Ralph G. (2012). "Reframing the Meaning of Self-Directed Learning: An Updated Model." *Adult Education Research Conference*. https://newprairiepress.org/aerc/2012/papers/ 22

Immordino-Yang, M. H. (2011). Implications of Affective and Social Neuroscience for Educational Theory. *Educational Philosophy and Theory*, 43(1), 98–103.

Katundu, M. (2019). Which road to decolonizing the curricula? Interrogating African higher education futures. *Geoforum*. Volume 115, 150–152.

Lave, J., & Wenger, E. (1991). *Situated Learning: Legitimate Peripheral Participation*. Cambridge University Press.

Nyamnjoh, F. B. (2012). 'Potted Plants in Greenhouses': A Critical Reflection on the Resilience of Colonial Education in Africa. *Journal of Asian and African Studies*, 47(2), 129–154.

Papert, S. (1980). *Mindstorms: Computers, children, and powerful ideas*. NY: Basic Books.

Resnick, M. (1998). Technologies for lifelong kindergarten. *Educational Technology Research and Development*, 46(4), 43–55.

Schwartz, D. L., Tsang, J. M., & Blair, K. P. (2016). *The ABCs of how we learn: 26 scientifically proven approaches, how they work, and when to use them*. W W Norton & Co.

Staab, L. (2020, June 2). Creating a project-based degree at a new university in Africa. *6th International Conference on Higher Education Advances* (HEAd'20).

Wenger, E. (2011). *Communities of practice: A brief introduction*.

Designing for Inclusion
Lessons from including all citizens

Jennifer Hardwick, Fiona Whittington-Walsh, Kya Bezanson, Anju Miller, and Emma Sawatzky

"Education as the practice of freedom affirms healthy self-esteem in students as it promotes their capacity to be aware and live consciously. It teaches them to reflect and act in ways that further self-actualization, rather than conformity to the status quo." – bell hooks (2003)

Today in British Columbia, there are few inclusive post-secondary academic options that treat students with intellectual and developmental disabilities as capable of living conscious, self-actualized lives on par with their peers. The majority of students with disabilities end up in community based segregated day programs or segregated adult special educational programs in universities (Rasmussen, 2002), where the opportunity to build agency and self-esteem in the ways that hooks envisions are limited. As Kya Bezanson notes, this discrimination is present throughout the education system:

> I am on the Board of Directors of Inclusion B.C and I am on the Self-Advocates Committee. I am a person with a disability. I have fetal alcohol syndrome disorder. Because of my disability a lot of my high school classes were modified, and I was told not to be proud of that work because it was modified.

Including All Citizens (IAC) was launched at Kwantlen Polytechnic University (KPU), on the shared, unceded and ancestral territories of the Coast Salish peoples, to provide an alternative to segregated or modified programs and to offer students with disabilities pathways to pursue their own unique self-actualization through education.

IAC works within a framework of disability rights to create an inclusive pedagogical model and is one of the first fully inclusive, for-credit university certificate programs in North America. More specifically, IAC has the United Nations Convention on the Rights of Persons with Disabilities (UNCRPD) as its philosophical foundation. Article 24, *Education*, recognizes the rights of people with disabilities to an inclusive education system including post-secondary, "without discrimination and on equal basis with others." It also recognizes that access to education is fundamental to "The development by persons with disabilities of their personality, talents and creativity, as well as their mental and physical abilities, to their fullest potential," and that education is necessary for "enabling persons with disabilities to participate effectively in a free society."

IAC was created through dialogue between Dr. Fiona Whittington-Walsh and five students with intellectual disabilities. The students had just completed KPU's adult special education program, *Access Program for People with Disabilities* (APPD) and were disappointed with the lack of opportunities that they were provided. IAC developed out of a qualitative, ethnographic, and participatory action research project that investigated and assessed teaching strategies and techniques that support all students in the learning of essential knowledge and skill sets. Research ethics was approved in 2016, and the pilot started with all five students taking courses on-par with their neurotypical peers. Without adapting the academic foundation and content of the curriculum, IAC uses the principles of Universal Design for Learning (UDL) to transform teaching and deliver curriculum to a wide range of learners. In essence, as Kya notes, "now, at KPU, my courses are not modified, and I take the class just like everyone else."

IAC is a student-centered learning environment where everyone is included and valued on an equal basis. The first three graduates (who are co-authors of this article, Kya Bezanson, Katie Miller (who uses the name Anju with friends and family), and Emma Sawatzky) completed their Faculty of Arts Certificates (FAC) in December 2020. The FAC pre-exists IAC and is an exit credential that prepares students for critical engagement with their communities. A new IAC cohort is scheduled to start in fall 2022.

Paulo Freire (1998) states that instructors need to "understand the concrete conditions" of student's lives and without this understanding, "we have no access to the way they think, so only with great difficulty can we perceive what and how they know" (p. 58). In order to truly understand the conditions of student's lives as Freire insists, one must develop relationships with students. Mentoring is a key aspect to IAC and is a two-fold process between instructor and student and between instructors. Mentoring involves an active relationship where individuals receive guidance and modeling that helps them enhance their professional growth and development (Cokley, 2000; Mertz, 2004; DeFreitas, 2007). Norma Mertz (2004) recognizes three functional categories in her hierarchy of mentoring framework that are associated with different roles: (1) modeling, (2) advising, and (3) brokering. (1) Modeling involves the roles of peers as well as teachers with psychosocial development.

It has well been documented that undergraduate students who have a relationship with a mentor experience a greater sense of belonging and connection to the university, which further facilitates academic success (Cokley, 2000; Lundberg & Schreiner, 2004). For students from marginalized and underrepresented groups, including those who have not excelled in academics, the relationship they develop with a faculty member becomes central to their learning and success (Stocks, Ramey, and Lazurus, 2004) superseding even the influence of their past experiences on their learning (Lundberg & Schreiner, 2004; Tillman, 2005).

IAC involves all three categories and subsequent roles. The instructor-student relationship is transferred into the classroom and provides the foundation for student-peer relationship building and mentorship. Relationship building becomes a key learning outcome and is an important aspect to IAC. (2) Advising is associated with professional development, and (3) brokering is associated with career advancement. Mentoring teachers have been identified as an important part of teacher education, recruitment, and retention (Wang & Odell, 2002; Tillman, 2005) and can become a "catalyst for transformative leadership" (Tillman, 2005: 614). Further, IAC offers what Wang and Odell (2002) call a *critical constructivist mentoring relationship* where together, the instructor/mentors are actively challenging existing teaching practices with the goal of teaching transformation. Instructors are mentored in how to transform their teaching and then become instructors/mentors for other faculty. This process has the potential for influencing wider systems change.

IAC's disability rights framework, and its core principles of mentoring, communication, and relationship-building are reflected in this collaborative, co-written chapter. Dr. Fiona Whittington-Walsh is IAC's principal investigator, coordinator, and mentor; Dr. Jennifer Hardwick is an instructor/mentor who began teaching with IAC in 2019; and Kya Bezanson, Emma Sawatzky, and Anju Miller are IAC alumni who recently graduated from KPU with a Faculty of Arts Certificate. We have brought our voices together in an informal and conversational way to explore different facets and experiences of Jennifer (Jen)'s Fall 2020 English 1202 course, which to-date is IAC's only fully digital offering. We hope to highlight not only how the course was designed and institutionally-supported, but how it was experienced by students who were learning online for the first time. We believe that 1202 demonstrates while flexible and thoughtful design matter, accessible online courses work best when they are grounded in transparency, individual and institutional collaboration, and strong relationships between faculty, staff, and students. Ultimately, we see inclusion as a collective responsibility and inclusive design as action-oriented, relational, and practical. We hope our discussion of 1202 acts as a case study in using critical digital pedagogies, universal design for learning, and open educational practices to collaboratively transform classrooms, institutions, and communities.

Context

Kwantlen Polytechnic University (KPU) is an ideal place for a pilot project like Including All Citizens (IAC). KPU is an open-access special purpose teaching university where faculty typically teach four courses each semester. While classes tend to be smaller (usually between 25 and 35 students), there are no teaching assistants, which means that faculty are the primary contact and support for up to 140 students each semester. Many of these students face linguistic, cultural, technological, financial, and personal barriers, which means that inclusive and accessible course design is imperative. Nearly 30% of the university's student population is composed of international students, of which 88% are English as an Additional Language (EAL) learners. Additionally, 32% of domestic and 48% of international students are first generation learners, and 35% of domestic students and 10% of international students identify as having a disability (KPU Office of Policy Analysis [OPA], 2021). Almost 70% of the student body works at least part-time while attending school (KPU OPA, 2022), and, like many other college students on the territories currently called North America, KPU students face significant challenges with food and housing security, homelessness, trauma, and discrimination based on factors such as race, gender, sex, and ability.

Given KPU's mandate and demographics, it should be no surprise that IAC's goals and approach have broad benefits. While the project was launched to support students with intellectual and/or developmental disabilities, placing access and inclusion at the forefront of course design benefits all students. However, figuring out what inclusive and accessible means has not always been easy. Ultimately, the answer for IAC has been about merging UDL and critical pedagogy, creating multiple pathways, and designing broadly while seeing students "in their particularity as individuals" and engaging with them "according to their needs" (hooks, p. 7).

Adding and Removing Barriers: Inclusive Digital Pedagogy

Drick Boyd (2016) outlines potential concerns and challenges to online learning from a Freirean pedagogical approach. Concerns regarding access to digital technologies are matched with fears that online learning is offered not because of pedagogical interests but for economic and political reasons. There are further concerns with the disembodied nature of online learning which could impede the constructivist nature of the learning process. Despite these concerns and challenges, Boyd does conclude that online learning can in fact, "enhance the teaching-learning experience" (p. 177) because of its ability to create meaningful dialogue as well as providing open access to information which can create opportunities of engagement for remote and marginalized communities.

Universal Design for Learning affirms that learners enter our classrooms with unique and diverse skills, challenges, and interests. As such, providing options and creating multiple pathways that all students can access is the best way to design an inclusive course. A digital environment does not change this core approach, and in our experiences, it does not make UDL any more or less difficult to apply—it simply changes the process. As IAC students so beautifully explain, digital pedagogy, like all other pedagogies, can simultaneously increase and remove barriers. For example, Anju struggled with a lack of social connection—particularly with the instructor—and with the mediated environment; however, she appreciated the ease of attending school from the comfort of home.

> School traditionally has been in person for most of everyone's life. But due to the Covid-19 pandemic, university students like myself have had to switch to online. The sudden switch has been quite difficult to adjust to due to several factors.
>
> One of the most difficult for me is that, as a visual learner, I require to be up close to the instructor. This helps me watch how an instructor relays their lessons (ie: gestures, pacing, individual eye contact, etc.). While watching the teacher's movements, I am able to retain more meaning behind each lecture. Most teachers had to resort to teaching online vocally, although some still did include visuals. This does help to aid in the visual part, but it doesn't always help give meaning behind their lecture.

> The physical interaction between students and teachers, and students and students, is still very much missing. We as people really depend on the social aspect of life, and to lose that just feels very foreign. This causes us as students to also rely on our computers and internet to get things done each session. During a particular lecture week, Moodle [KPU's Online Learning Management System], was down and we were all unable to do the [synchronous] class that week. That caused a shift in the schedule moving forward.
>
> While having no physical interaction was extremely difficult, there were plus sides to classes being online. The first has to be the convenience that you could quite literally jump out of bed (because the instructor was the only one with video) and join the class, instead of having to get up extremely early and take both the public and school transport to get to school. I did not have to leave my house almost 3 hours before school to be there on time.

Emma missed the focus and social interaction of classroom work, but appreciated the flexibility of having both asynchronous and synchronous options for learning in an online class.

> What I didn't like about online class was that I didn't get to interact with people. Also, I didn't really have technology problems; it was more like interruptions. Lots of people at home interrupting me because everyone is at home right now instead of going to work.
>
> What I liked about the online course is how flexible it was.

The students' experiences throughout the course affirmed that a digital environment was not inherently more or less accessible for all students—it simply offered different challenges and possibilities.

Where Open and Critical Pedagogies Meet: Course Materials

When designing an online version of English 1202—which is a writing-intensive, mandatory first year literary studies course—Jenn drew from Universal Design for Learning, critical pedagogy, and open education practices in order to reduce barriers, foster community, and provide a balance of structure and choice. As previously stated, mentorship is key to IAC, and Fiona mentored Jen in inclusive pedagogies by providing input into the course design, liaising with the library to source accessible materials, and collaborating on assignment design. Throughout the course, Jen and Fiona regularly checked in with each other, and Jen regularly checked in with all 1202 students through informal forum questions, surveys, and class dialogues.

Flexibility is absolutely critical to UDL, so the course was built with multiple pathways; students could complete 1202 entirely asynchronously or partake in synchronous learning and community activities. While UDL is primarily concerned with structure and design rather than content, critical pedagogy reminds us of the importance of curricula that challenge systems of power and provide students with tools to transform the world. Jen's goal—to borrow from bell hooks—was to use both UDL and critical pedagogy to "enact pedagogical practices that engage directly both the concern for interrogating biases in curricula that reinscribe systems of domination (such as racism and sexism) while simultaneously providing new ways to teach diverse groups of students" (hook, 1994, p. 10). The course theme was "Literatures of Resistance" and the texts were chosen to speak to the diverse lived experiences of students and encourage them to meaningfully and ethically engage with communities they may not be part of. The texts—which included multimedia videos, spoken word, personal essays, a novel, and short stories that spanned genres and mediums—centered the voices and lived experiences of Indigenous, Black, disabled, women, LGBTS2S+, and South Asian authors and directly addressed topics such as colonization, racism, gender-based violence, and ableism.

As Anju notes, having a diversity of voices made the course more engaging for students. The texts spoke to the lived realities of the classroom community, modeled different forms of individual and communal agency and resistance, and provided opportunities to reflect on topics that felt meaningful:

> Every [text] that was assigned became extremely relatable. Each one varied to show different experiences, which made it memorable. Some examples that were touched upon [were] gender, sexuality, and disability. This made doing the homework more enjoyable.

Kya echoes Anju, noting that course texts and assignment options provided her with opportunities to connect her lived experiences as an

Indigenous woman with the course content and explore topics such as colonization, resistance, community, and culture.

> My favorite part of the course was having to read, *The Marrow Thieves* [by Cherie Dimaline]. A lot of the work I chose to do was about Indigenous people. I just loved that book and the music video, "March March" by the Dixie Chicks. These two [texts] really were the ones that spoke out on a lot of issues that we are facing today. Especially during the pandemic. And a lot of issues that [Indigenous] people have been facing for centuries.
>
> I am super passionate about the [Indigenous] community so I am glad [Jen] let us choose the topic for our final essay. I am passionate about this topic because I am Indigenous. Learning about my culture is really important to me. I felt like I was learning about my culture while reading the novel, and that is important to me.

Emma also found learning about Indigenous communities powerful as a non-Indigenous woman, and related to *The Marrow Thieves*' emphasis on family, community, and respect for different ways of knowing.

> The novel is now one of my favorites. I could relate to the story, despite not being Indigenous. I found the story was about building a strong team and community, creating strong connections. Everyone in the community was valued because of the different things they could contribute. For example, [the character] Minerva was a senior and even though she could not fight and do combat she was valued because she was a knowledge keeper and a storyteller. So, everyone kept her safe. When one person struggles, everyone joins together to help them. Just like in my family. So, I could relate to the story a lot.

The diversity of stories, voices, and approaches present in the class content and conversation drew attention to the strengths of different ways of knowing, learning, and sharing. It also helped students develop intersectional understandings of power as they realized there are many overlapping forms of both oppression and resistance. Given that IAC is about challenging systemic discrimination and transforming educational practice, it was important that the curriculum align with instructional design framework.

One key aspect of providing accessible course material for students is audio recordings to accompany required readings. In previous English courses, Fiona recorded numerous short stories and articles for Jen. For English 1202, we wanted to expand our use of audio readings and engage with an audio recorded novel. *The Marrow Thieves* by Cherie Dimaline (2017) was the assigned novel and is available in print, e-book, and audio book. We decided to maximize UDL principles by offering all three versions. Students had the choice of what version they would engage with.

We were also committed to KPU's "ZTC" initiative, which is Canada's first Zero Textbook Cost program. Currently, there are over 800 courses that use open educational resources and/or library materials and several program degrees, including the Faculty of Arts Certificate. We didn't just want to suggest to students that they could purchase the audio recording themselves; we wanted to provide all three of the available mediums for the novel as part of the open education initiative for no cost to the students. In order to be able to do all of this, we enlisted the support from KPU's E-Resource librarian, Allison Richardson, who explored every aspect of adopting all versions of the novel as library resources. This was new terrain for both the IAC team and the library, but we were all committed to full accessibility. Allison discovered numerous issues with adopting the audio book. We had to use OverDrive as the medium to host the E-book and audio book. OverDrive is a free service for students offered by libraries that provides digital content such as ebooks and audiobooks. KPU had to purchase a one-time subscription to OverDrive for 1202. Students had three ways they could access OverDrive: Libby app on a phone or device; Libby website on a desktop/laptop; and on the OverDrive website. Students could also download the ebook onto a kindle.

Despite providing links to the ebook and audiobook on the 1202 Moodle website, students still had to set up individual OverDrive accounts to access the content. Because students were choosing which medium they wanted to engage with, this eliminated any concerns regarding privacy of personal information. Students were not being forced to set up an OverDrive account and provide personal information, it was their choice to do so.

One other concern we encountered was that we could not provide all versions to all 25 students. The ebook had a different model with unlimited simultaneous users, so there wasn't a need to purchase individual copies. The audiobook, however, was based on a pay-per-individual use. We decided to order five print copies for the library, unlimited ebooks, and ten copies of the audiobook. The KPU library supported this initiative and covered the costs. In moving forward, we are trying to secure a permanent subscription to audiobooks to accompany print and ebooks.

While the media were not without technical issues (such as not being able to sign out multiple versions at the same time), students enjoyed accessing the novel via the different media. Emma liked the audiobook and found that it supported her learning needs in a way that print book could not.

> There was also an audio recording of the novel, *The Marrow Thieves*. I just got to listen to it and it was much easier for me to understand rather than just reading it, because I have reading comprehension problems.

Designing for Clarity and Community

English 1202 was an asynchronous course with optional synchronous components. It was designed from the ground up in order to promote accessibility, transparency, instructor presence, student participation, and community. Our Moodle website provided weekly schedules and acted as a central repository for documents and resources. Students cannot become part of a learning community if they cannot access that community, so intentional, accessible design was at the very heart of how ENGL 1202 was set up. We endeavored to reduce cognitive overload by avoiding clutter and keeping text spaced out so that students could navigate and find relevant information. Headings, labels, and hyperlinks were descriptive to ensure that they could be read by screen readers, all images included alt-text, and all videos contained captions. As Whittington-Walsh (2021) concludes:

> "It is recommended to utilize a fully accessible course website. This provides the course information in multiple accessible and portable mediums. Conceptualize your course website as the more detailed road map that can store numerous documents uploaded for additional information. It can be set up the same way as the syllabus: not overly cluttered with text and nicely spaced out so the reader can navigate and find the relevant information including week-by-week schedule, homework, assignments, and grades. Larger font size further benefits all learners. Including a to-do list for each week is also very beneficial for students."

Readings, resources for learning and well-being, and a week-by-week schedule were available on the site a week before the class began, and then the individual weeks were populated with additional materials as the semester progressed. Every Wednesday morning a 20-40 minute asynchronous lecture was posted, alongside the lecture slides, a forum with discussion questions, and a checklist that outlined the weekly activities such as readings, deadlines, and workshops. While some of the students struggled with the transition online, they found the consistent and clearly-labeled format easy to navigate. Anju noted that the headings and use of color made the class schedule easier to follow.

> The syllabus was extremely easy to follow. The use of diverse colors that represented due dates, assignments, online readings, and class discussions helped keep me focused.

Kya also appreciated the clarity and structure of the online environment.

> Jen's online class was really well-organized. I didn't need to look at the actual syllabus because her online format made it super easy to find things.

The visual navigability of the learning management system was a huge benefit of online learning, and clearly marked resources, weekly checklists, and a structured approach to posting helped to keep students on track.

Engaging with community online was central to the 1202 learning experience, and all 25 students in the course did it a little bit differently, bringing their unique preferences, approaches, challenges, and perspectives. On Wednesday afternoons students had the option to attend a Coffee Hour Chat, which functioned as a space for informal discussion about course content. Both the Coffee Hour Chat and the forums counted towards the course's participation grade (which also included activities and workshops as possibilities), which gave students choice in how they wanted to share their knowledge and engage with the classroom community. Synchronous workshops were also offered throughout the semester to help students work on specific skills such as writing, editing, and library research. The workshops were interactive, and they were recorded for those who couldn't attend. Jen acted as a facilitator in both Coffee Hour sessions and the forums by asking questions and helping students make connections between their responses, but the conversation was nearly entirely driven by students. Similarly, while Jen set the topics for the workshops and guided peer-review sessions, they were designed based on students' interests, questions and requests. In addition to providing space to discuss course content, build skills, and foster relationships, Coffee Hour Chats, forums, workshops, and additional activities such as surveys allowed students to regularly provide feedback to Jen about deadlines, assignments, discussion questions, and readings.

Offering multiple synchronous and asynchronous ways to engage with the classroom community proved to be integral to student well-being and success. All of the students participated in different ways, and made use of different spaces and tools. Emma liked having options but gravitated towards the synchronous Coffee Hour Chats.

> Jen gave you a choice for how you participated. Coffee Hour was people talking about anything, like "what's going on?", or "I am struggling with this section. Can you guys help me?" The workshops we had to follow specific guidelines: we are talking about this and nothing else. The workshop was more like sitting in her class except we were all on screens. What we were going to talk about in the workshops was on the syllabus, so we knew before what was going to be discussed.

> There were also forum posts. Jen would post questions and you would answer the questions if you could not attend the Coffee Hour or the workshop and that's how you got participation marks. I went to every Coffee Hour and the workshops. I didn't find the forums interesting. The Coffee Hours were easy marks! You come and talk with Jen and other classmates. The workshops were great because you knew what the topic was ahead of time, and they were recorded. So, if I didn't understand something I would listen to it again, and then I understood it.

Anju also attended the Coffee Hour Chats and found that the multimodal dialogue (which included chat, video, and audio functions) was easier for her than in-person interaction.

> During class discussions I was able to find talking to my classmates much more alleviating. This is due to many years meeting most of my friends online in video games. I have found that talking online over in-person had many pros and cons; we as humans need to have the physical interactions with one another to function, but personally, I communicate better online than in person. The difficult part of communicating online is after the semester I don't maintain the relationships like I would have

in person. This is because physical interactions allow for easier in-person relationships to build [...] written communication won't grow [relationships] as fast.

On the other hand, Kya found that the asynchronous forums often worked better for her, but she shifted between options depending on the week.

I did the forums and mostly the workshops and a few Coffee Chats to help me understand the class work more. I really liked that there were choices for everything.

Having multiple ways to engage with community, build relationships, and learn from one another helped reduce isolation and foster a broad community of support where students could look to each other for help and encouragement.

Assessment and Feedback

UDL frameworks recognize that "there is not one means of action and expression that will be optimal for all learners." As such, it is recommended that UDL practitioners provide students with multiple ways to share their knowledge. This was not always easy in English 1202, which is a writing-intensive course with fairly strict learning outcomes. According to the course outline students must:

- Apply the writing process in writing about literature
- Write analytical essays about literature that develop and defend a clear, substantive thesis
- Write literary analysis using correct, clear, coherent, and effective English
- Apply basic research techniques by using secondary sources responsibly
- Write successfully under time restrictions

There is no way for a student to succeed in 1202 without writing and editing academic essays, which means that the available modes of expression are somewhat limited.

Applying a UDL lens to a writing-intensive course required looking at the course as a whole; while not every assignment could contain options for expression, options could be provided throughout the semester. Additionally, writing assignments could offer choices in topic and approach, which would allow students to focus on materials they were passionate about.

Throughout the semester students were given assignments that asked them to:

- Write or verbally present an essay proposal
- Write analytical and research essays
- Take short open-book quizzes to build vocabulary and show reading comprehension
- Participate in forums or Coffee Hour to practice literary analysis skills, collegial engagement (for example building off of others' ideas or engaging in respectful critique), and summary

Students were also invited to engage in peer-editing workshops and self-assessments/reflections at different points in the semester. In addition to building writing skills, the assignments were designed to give students the opportunity to share knowledge in ways they found comfortable, and to encourage them to try new or challenging forms of expression.

The students appreciated the variety of assignments and topics available to them, and, as expected, found some forms of assessment more difficult than others. However, what they found engaging, challenging, easy, or boring differed drastically according to their interests, skills, and barriers. While this may seem obvious, it is worthwhile to note that no assignment was inherently more or less accessible to all students—each form of assessment reduced barriers for some students, while increasing them for others. As a result, it was important to acknowledge obstacles, provide resources, and create pathways so that students could navigate through and around them.

For example, Kya struggled with vocabulary quizzes, but found reading comprehension quizzes and essays easier and more engaging.

> There were two tests and I hate tests. One was on vocabulary and the other one was on the novel. I studied really hard for the vocabulary test, but I still got really stressed. I can't associate things when it's vocabulary like I can for novels. I remember characters because I create them in my imagination when I am

reading a novel, but you can't do that for definition questions. We also had to write essays for other things. We could choose [topics]: pick one of the [stories], poems, or music videos and write about it. "What are they trying to get you to understand?" kind of thing.

Anju also found quizzes stressful, but quickly realized that making an office hour appointment to prepare was helpful. She also found that the format of online quizzes alleviated anxiety and helped her focus.

> When things like quizzes come up, I personally have high anxiety even when I know the material. My instructor offered me the ability to talk with them over Moodle, which in turn helped de-stress and center me to take the quiz with confidence. While taking the quiz online, I find having one question on the screen at a time helps calm my nerves. Physical classes don't have the luxury to give out one question at a time and then let you revisit them later on before submitting.

Anju also appreciated that assignment instructions were offered in multiple ways (usually through a written assignment sheet and accompanying video) and that she could draw from different kinds of texts and sources in her assignments.

> This made doing the assignments after much easier to follow thanks to the online videos posted. The assignments also varied greatly from having to read a book, to watching and listening to videos on YouTube that corresponded to the assignments.

Verbal and written formative and summative feedback were provided throughout the course. The feedback was holistic and relationship-oriented in that it addressed students' growth, provided resources, and offered encouragement in addition to commenting on students' abilities to meet learning outcomes and assessment criteria. As the students note below, the relationship to the instructor was instrumental in the feedback process. The online environment made comments feel disembodied, so personal touches were significant. Trust, communication, and transparency facilitated learning and problem solving, and gave students the self-confidence to meet challenges.

Anju made regular use of online office hours and sought consistent feedback, but found the lack of physical presence a challenge.

> Homework has always been very difficult for me since I heavily rely on the feedback from my teachers and therefore, put myself down if I don't get enough reassurance about how my work could be improved constructively [...] For example, having to type out something and then seeing a curser edit for you over a physical being makes it feel like a robot over a person is editing your work.

Kya also missed the physical presence of the instructor, and valued personal communication alongside the more standard feedback.

> Another part of online courses is you can't see the teacher's enthusiasm with your work like when they hand it back to you in-person. Even though my grades were the same, I still missed having my assignments handed back to me and getting that compliment from the teacher. That always made it worthwhile and made you feel like all that hard work was worth it. Online courses you hand it in and get your grade. I also worried that she might not have gotten it. When you physically hand in something, you know they got it.

> At the end of this online course, Jen sent me a little message saying how well I did and that made me feel really proud. To see that message from Jen — it just put me over the top! I couldn't believe that she thinks this, but she does. That was important for me to get her message.

The students' experiences highlight the importance of ongoing communication, and of feedback that acknowledges students individually. Assessment is not just about guiding learning—it is about building relationships.

Conclusion

The success of IAC is grounded in critical digital pedagogy, universal design for learning, and open education approaches that ask us to

- treat accessibility as foundational and necessary for all learners;
- acknowledge students as complex, whole people whose learning is impacted by social, political, and economic structures;

- forefront communication, community, and relationality;
- promote reflection, action, and agency;
- acknowledges, upholds, and celebrates diverse voices and methodologies;
- build networks of support individually and institutionally; and
- create multiple pathways so that students can engage with knowledge in a variety of ways.

We hope this case study of 1202 shows the multifaceted nature of inclusive instructional design. It is work that is grounded in theory, care, community, and an attention to detail. Most significantly, it should not be siloed or seen as the responsibility of one individual. Inclusive instructional design is at its best when it is approached collaboratively; faculty mentorship, ongoing consultation with students, widespread student supports, and collaboration between units (for example, with the library) have a significant impact on its success.

Students with intellectual and/or developmental disabilities deserve equal access to education. IAC proves that it can be done, and that inclusive, for-credit academic options are possible. What is needed is willingness, time, and commitment, both on the part of instructors and on the part of institutions. If we are truly committed to education as a transformational, transgressive practice of freedom, then inclusive instructional design is a necessary starting place.

References

Bozeman, B. & Feeney, M.K. (2007). Toward a useful theory of mentoring: A conceptual analysis and critique. *Administration & Society*, 39 (6), 719 – 739.

Cokley, K. (2000). Perceived faculty encouragement and its influence on college students. *Journal of College Student Development*, 41(3), 348-352.

DeFreitas, S. C. & Bravo, A. (2012). The influence of involvement with faculty and mentoring on self-efficacy and academic achievement of African American and Latino college students. *Journal of the Scholarship of Teaching and Learning*, 12(4), 1–11.

Friere, Paulo. (1998). *Teachers as Cultural Workers: Letters to those who dare to teach*. Westview Press.

hooks, bell. (2003). *Teaching community: A pedagogy of hope*. Routledge.

hooks, bell. (1994). *Teaching to Transgress: Education as the Practice of Freedom.* Routledge.

KPU Office of Planning and Accountability (2021). *KPU Student Profile.* https://www.kpu.ca/sites/default/files/Institutional%20Analysis%20and%20Planning/KPU%20Student%20Profile%20-%20April%2013%202021.pdf

Lundberg, C. A., & Schreiner, L. A. (2004). Quality and Frequency of Faculty-Student Interaction as Predictors of Learning: An Analysis by Student Race/Ethnicity. *Journal of College Student Development,* 45(5), 549–565. doi:10.1353/csd.2004.0061.

Mertz, Norma T. (2004). What's a Mentor, Anyway? *Educational Administration Quarterly.* Vol. 40, No. 4, 541–560

Prince, Michael. (2009). *Absent Citizens: Disability Politics and Policy in Canada.* University of Toronto Press.

Rasmussen, L., Haggith, K., & Roberts, J. (2002) Transition to Adulthood, Moving Needs IntPractice: A Canadian Community Partnership Response to New Adult Service Needs for Individuals with Disabilities. *Relation Child and Youth Care Practice,* Vol 25 (3), 29–38.

Stocks, J., Ramey, J., & Lazarus, B. (2004). Involving faculty at research institutions in undergraduate research. In L. Kauffman & J. Stocks (Eds.) *Reinvigorating the undergraduate experience: Successful models supported by NSF's AIRE/RAIRE program.* Washington, DC: Council on Undergraduate Research, 7–8.

Tillman, Linda C. (2005). Mentoring New Teachers: Implications for Leadership Practice in an Urban School. *Educational Administration Quarterly,* Vol. 41, No. 4 (October 2005), 609–629.

United Nations. (2006). *Convention on the Rights of Persons with Disabilities.* Treaty Series, 2515, 3

Wang, J., & Odell, S. (2002). Mentored learning to teach according to standards-based reform: A critical review. *Review of Educational Research,* 72(3), 481–546.

Whittington-Walsh, F. (2022 forthcoming). Universal Design for Learning and Inclusive Post-Secondary Pedagogy: The Including All Citizens Project. In Frederic Fovet (Ed.), *Third Pan-Canadian Conference on UDL: Conference Proceedings.* Fovet, Frederic (Ed). EdTech Books.

The Straight and Narrow is the Path of Least Resistance, and I Believe, We Need Resistance, or at Least, We Need to Nuke Las Vegas First

Pat Lockley

		In this piece,	
negative			
		words appear on the left, and	
			positive
		words appear on the right	
		This isn't for political wings, more just mostly because it looks	
			pretty
		and makes sure that i'm being really, really clear, because	
			fuck
nuance.[1]			
Nuance			
		now	
			fucked
		and with	
			apologies
.			

---

1. https://journals.sagepub.com/doi/abs/10.1177/0735275117709046?journal-Code=stxa

	for bad language aside, everything we do in design is a structure, or a form, or these days, a pattern. Patterns, and that if we are to be critical, then we must criticise our words, letters and full stops You can argue three columns is not enough range, or too simple, but then when your	
LMS		
	only gives you	
three options		
	that is sufficient? Elearning	
		sword-pens
	bent into	
		ploughshares
	is a victory purely for the blacksmiths of this world. If	
		Domain of One's Own
	is a pattern, then that is a pattern based on moving the voice of the designer to the fore, not	
heating and hammering it,		
	via a series of	
dropdowns and checkboxes		
	like the doomed child of	
data entry forms		

	and	
		minesweeper.
	If we aim to design with tools that grant us more	
		agency,
	do we need to question the role of ourselves in this process? If we are critical of instructional design, do we also need to be critical of our institutions, do we need to know where our criticism stops? We live in a world of universities who can't rename buildings named after	
undoubted racists		
	because to do so would be to	
"rewrite history,"		
	but those same universities require a new	
$100 textbook each year		
	because, apparently history can be rewritten?	
		¯_(ツ)_/¯
	So are we really "rewriting history"? Can you imagine a timeline in which Oxford's LMS is called	
Rhodes?		

	Because it's both clever and you know, a	
		pun.
	We know	
Negative stereotypes[2] worsen education,		
	this remains unaddressed, we know students of colour do not progress through to Masters courses[3]. We maintain the contradiction of being a site of learning, whilst not demonstrating what we have learned. Or perhaps, reader, we do know. Perhaps we aren't concerned with history, but with place. Universities as a machine of place, universities as	
		notaries
not–aryaning.		
	No one can be a	
gatekeeper,		
	unless someone has	
built a wall.		

---

2. https://www.psychologytoday.com/gb/blog/choke/201008/when-negative-gender-stereotypes-hang-heavy-in-the-classroom-girls-learn-less

3. https://www.universitiesuk.ac.uk/policy-and-analysis/reports/Documents/2019/bame-student-attainment-uk-universities-closing-the-gap.pdf

	The place offered to a student becomes a place in society via the university's prestige. The pattern of becoming a Harvard man or a Yale man is the pattern of being a grade. We've spent the last year with		
		medical grade,	
	and we've always had		
weapons-grade			
	and		
making the grade.			
	The grade is not made by the student, it is made for the student, unless we choose a pattern without it, or maybe, critically, a pattern with it, but without its power. If we chose,		
		ungrading	
	explicitly is the course seen as lesser, what if instead, as our pattern, we just choose to release everything to ensure the best grade possible. I know,		
		dear reader,	
	you worry about it. What will		
		teen dramas	

	do if stripped of a storyline?	
		Reader,
	do not worry, Netflix has so many these days, that losing one story line only aids the rapid	
burning of venture capital make rocket go vroom vroom money[4]		
	What of the universities? Well,	
dropout		
	re-appeared during the great	
MOOC, avalanches, tsunamis, or insert natural disaster of choice here,		
	But now we teach our	
income streams		
	—sorry—	
		students
	to be resilient. We jest	
		jesters
	over burning through money, but with students the money that's burning is theirs. So what is the pattern we need? Elearning remains obsessed with the new. A	
FOMO		

---

4. https://allpoetry.com/poem/8593429-Rocket-by-Edvard-Kocbek

	of trends and not trend-settings. Of being a	
perpetual thought leader,		
	of zeitgeist and not	
shite guess.		
	It remains only a matter of time before the horizon (and the	
Horizon report)		
	gains a new shadow.	
Flipped learning, gamification, mobile learning, serious games		
	What do they all share? Taking time from students. Flipped learning is increasing the workload for a student, gamification and serious games are making the learning more addictive, mobile learning requires an "always on" mode. None of this was moved into column as I want you	
		dear reader
	to do it.	
	We know students are	
poor,		

	but we need to understand what they are poor of. How many poverties does a person need to have before a change happens. Here's another list	
Time poor, Financially poor, Opportunity poor.		
	We can build a beautiful pattern, our elearning ink mellifluously making courses, but with the students as worker bees making this	
stolen honey		
	So which pattern? The anti-pattern[5] We can build a course to	
		inspire or encourage wider reading,
	but if we know they are time-poor students, then what pressure is that? On a day to day basis, our students may be more concerned with their	
gig economy star rating		
	to worry about their university grade. We need wider reading because	

---

5. https://en.wikipedia.org/wiki/Anti-pattern

"We are training them for jobs that don't exist yet"		
	Well, here's some news,	
those jobs may never exist		
	Is the anti-pattern minimalism, or is it friendship? Is it helping by freeing people to choose to spend what time they have how they want to or need to? So is giving everyone the best grade bad? No. Grains of sand do many things. They make	
		beaches to relax on, glass bottles to make molotovs
	and in the case of the anti-pattern of grading highly, the grain of sand does too beautiful things. One, it is the grain of sand which makes the	
		pearl,
	the	
		pearl
	which is very much the	
		oysters and not yours,

	and it is the grain of sand wearing down the	
teeth of the gears of the machine.		
	Which machine? Well I can promise you that	
algorithms		
	are already hunting through LMSs looking for patterns which lead to good grades. Imagine the machine learning that one pattern gets every student the best grade. Imagine that machine then recommending that as the best pattern. Remember, the machine learns that the only way to win	
thermonuclear war		
	is	
not to play.		
	So how do we win thermonuclear education?	
		By choosing to play differently, by choosing the move no one expected, by understanding that you need to play, but that perhaps you're playing with things that need playing with[6]

---

6. https://www.theguardian.com/sport/2021/mar/18/bongcloud-meme-opening-carlsen-nakamura

Access Alone Isn't Enough
Understanding and closing the college digital divide

Benjamin D. Remillard

If we are going to learn from our teaching and learning experiences during the pandemic, we have to understand the socio-economic forces behind the digital divide and how it will continue to affect our classrooms in the coming decades. This chapter will show that despite the tenacity of these issues, we can begin to implement holistic digital pedagogies at the classroom level that recognize, assess, and address the opportunity gaps our students face. While we may not have the power from the confines of our lecture halls to overthrow the systems that lead to these inequalities, we can narrow some of those divides by taking small, learner-centered steps toward making technology more accessible and impactful for our students.

America's Digital Divide

To better understand the root of the issues we see at the college level, we must understand how various gaps begin when students are still enrolled in primary and secondary school. These trends are easy to track using standardized tests, where wealthy students consistently outscore their peers in lower income brackets (Zumbrun, 2014). University of Connecticut researchers have shown that when it comes to the online research and comprehension gap between the richest and poorest students, by the time those students get to middle school there is already a year's worth of learning differences separating those groups (Braverman, 2016). Students of color are especially hard hit by this economic divide (Dixon-Román et al., 2013). When it comes to specific skills like reading and math, a 2016 report in the New York Times noted that, "Children in the school districts with the highest concentrations of poverty score an average of more than four grade levels below children in the richest districts," disproportionately affecting students of color (Rich et al., 2016). These disparities grow as students age.

This all relates to what sociologist Robert K. Merton coined the Matthew Effect, the basic principle of which is that the rich get richer and the poor get poorer. A similar principle applies to technology. Lloyd Morrisett, a founder of the Children's Television Workshop, first coined the term "digital divide" in 1996 "to describe the chasm that purportedly separates information technology (IT) haves from have-nots," where the gap between them creates inequality (Eubanks, 2011). As Neil Selwyn argues, technology reinforces the Matthew Effect's impact on the digital divide (cited in Braverman, 2016). Here is where gaps will develop, where "high-achieving" students build on previously developed knowledge and skills while the "lower-achieving" students potentially flounder due to lack of access, lack of education, or a combination of both. Focusing on achievement alone, however, misses the bigger issue.

Educators are shifting away from understanding these divisions as "achievements gaps" by understanding them instead as "opportunity gaps." Through the former lens, the differences in educational outcomes between students and school districts are often attributed to deficient work ethics, lack of "grit," etc. The latter lens, however, encourages us to understand the inequitable access to educational tools and experiences marginalized students face. This shift in terminology takes the onus off the students by focusing instead on the unbalanced systems that discriminate against those students in the first place (Mooney, 2018). Simply increasing access to technology, however, will not solve for these opportunity gaps.

Increased Access to the Internet ≠ Equitable Access to Opportunities

A 2013 report by the U.S. Census Bureau noted that smartphones were reducing the internet-use-gap between different racial and ethnic groups. This trend had been building for a few years when, during the last quarter of 2010, smartphones outsold PCs for the first time (Smart Phones Outsell PCs For The First Time, 2011). This has served as a cost-effective crutch in education. As noted by NPR in 2016, a quarter of low-income families with school-aged children had, "only a mobile device for internet access. Among families living below the poverty level the proportion rises to a third. And it's highest at 41% among immigrant Hispanic families in particular" (Selyukh, 2016). This begs the question: how can you complete all your school assignments with just a smartphone, especially if it's not a high-quality smartphone?

There is a concern in many low-income school districts that if you require students to complete homework online or require them to use specific computer programs, not all of their students have adequate access to that technology at home. Prior to the pandemic roughly 17% of primary and secondary school students in the US did not have access to computers at home, and 18% did not have access to at-home broadband internet (with expense being cited as the main reason for a third of those) (Associated Press, 2019). As is often studied, this divide is most apparent when examining access between rural and urban residents. In 2018, for instance, roughly 51.6% of rural US residents had 250/25 megabits per second internet access, while 94% of urban residents had the same (Lai & Widmar, 2021). Even in urban spaces though there can be a serious divide between the digital haves and have-nots. A recent paper argues that digital divide can at least be partially attributed to 20th century housing policies where, "the exclusionary property ownership and residential segregation technologies developed by administrators of the HOLC [Home Owners' Loan Corporation] and the FHA [Federal Housing Administration]" contribute to "current differences in students' geographies of opportunity as they relate to broadband access" (Skinner, et. al. 2021).[1]

As more attention has been paid to what has increasingly been referred to as digital redlining, so too has more evidence of it and its effects have been brought to light. As an example, a 2017 analysis of FCC data regarding AT&T's digital infrastructure plans "strongly suggest [that AT&T] has systematically discriminated against lower-income Cleveland neighborhoods in its deployment of home internet and video technologies" for over a decade. This includes the company's decision to withhold fiber-enhanced broadband from "most Cleveland neighborhoods with high poverty rates." This had a double effect whereby residents in those lower-income neighborhoods were not allowed to apply for AT&T's "Access" discount rate program because they lacked the minimum download speed required for the program. This then required those residents to rely more heavily on expensive (and slower) mobile broadband (Stanton, 2017). Such disparities, especially in areas with scant preexisting digital infrastructure and/or competition between service providers, limit students' opportunities while reinforcing the digital divide (Lai & Widmar, 2021).

From a college educator's perspective, we see this struggle perhaps most often among our commuter and community college students who don't always have 24/7 access to wi-fi like what is available for stu-

---

1. The authors proceed to note that "While some racial/ethnic groups and those with higher household incomes have greater broadband access overall, we find that otherwise demographically similar persons with the same ISP options may nevertheless have different likelihoods of in-home broadband access due to neighborhood characteristics that are correlated with New Deal-era housing policy," and that, "As our results show, even persons from the same racial, ethnic, or income group had different likelihoods of broadband access in connection with the historical rating of their neighborhood," 25, 27.

dents living in residence halls. As noted by NPR, if low-income students do not have wi-fi at home, or if they are on a metered plan, they are going to run into roadblocks (Selyukh, 2016). Even if those students can access a coffee shop or public library where they can use free wi-fi and maybe even a cheap printer, those places eventually close, shutting off access to those needed resources. Despite increased access to the internet and mobile technologies, then, the historically rooted gaps in accessibility continue to separate our students (Cortez, 2017).

While access to online services and digital technology improved during the first year of the pandemic as schools and governments scrambled to ensure continued learning, the fact remains that millions of students remained without reliable access to computers and/or wi-fi (Richars et al., 2021). How schools dealt with that transition varied, again, along socio-economic and racial lines.[2] While we can probably guess what impact this will have, the pandemic's long-term effects on education may not be fully understood for years to come. As educators, we can't solve these systemic issues on our own. We can, however, avoid perpetuating these inequities by creating meaningful learning experiences for our students that address these preexisting opportunity gaps and the resulting disparity in skills.

The Importance of Digital Literacy in a Digitally Divided Age

In *Pedagogy of the Oppressed*, Paulo Freire asked, how do you create a learning environment that does not simply educate, but empower learners? One of Freire's most important contributions is the notion that education should not be a top-down, sage-on-a-stage approach where we simply lecture at our students about various topics—otherwise known as the banking system of knowledge—because they need to meet government or university standards/learning outcomes. In high school or college-level classes, this is like making them learn how to use Excel, or a research database, or how to read a scientific article, without helping them understand how elements of those experiences are applicable in other parts of their lives. Freire instead endorsed the empowerment of students by working with them through an educational experience rooted in dialogue, by creating a system that works to their benefit (Freire, 1970).

---

2. For early analysis of those trends, see Emma Dorn, Bryan Hancock, Jimmy Sarakatsannis, and Ellen Viruleg, "COVID-19 and student learning in the United States: The hurt could last a lifetime," June 1, 2020.

This process will look different when taking into consideration the needs of different students, educators, and institutions. To borrow from a different writer while keeping in this vein, Sumun L. Pendakur advocates for educators to adopt, "a reflective, identity-conscious framework that grapples with the hegemonic nature of power" both in and outside of our particular universities. As Pendakur suggests, we need to become "empowerment agents" who better support students on the margins and their particular needs (Pendakur, 2016). It is on us as educators, then, to understand how opportunity gaps affect our students, to identify what gaps need bridging, and then to give them the tools necessary to create the desired changes they wish to see in their lives based on their own decisions, input, and professional and personal interests.

Many colleges have implemented 1:1 laptop, computer, or iPad programs to remedy any gaps they see in their technical offerings, by making sure all students have access to the technology they need (EdTech Staff, 2021). But we often forget that we still need to teach students how to best use those devices if those tools are going to make a difference in their lives and education; access alone isn't enough if they are going to use those devices to play games or watch Netflix (Eubanks, 2011).[3] This increase in accessibility also means little if faculty are not being trained or incentivized to incorporate these tools into their curricula.[4] As Sean Michael Morris has suggested, when we consider what critical instruction design looks like, "digital tools, strategies, and best practices are a red herring in digital learning" instead of asking "What tool will we need?," we need to instead ask, "What behaviors will need to be in place?" (Seam Michael Morris, 2018b).

To begin, we need to be clearer about how students do and don't use technology, especially at the college level, where we often assume students are more capable and technologically savvy than they actually are (Braverman, 2016). We often talk about students being on their phones in the middle of our classes, whether it's on TikTok, fantasy football, a mobile game, etc. What this shows is that they are mobily literate, which is "the ability to navigate, interpret information from, contribute information to, and communicate through the mobile internet, including an ability to orient oneself in the space of the internet of things (where information from real-world objects is integrated into the net) and augmented reality (where web-based information is overlaid on

---

3 The Netflix example comes from my own anecdotal experiences in and out of the classroom.

4 For an example of an institutionally required faculty training, see Craig Guillot, "Bridging the Technology Gap in Higher Ed," March 10, 2021.

the real world)" (Hockly et al., 2014). That is, students who use their phones a lot are good at using their phones.

Having a good phone, though, rarely helps students when it comes to learning the higher-level digital skills we expect them to develop for college—whether it be writing a report and formatting it using a word processor, doing digital research, pulling/analyzing/synthesizing information from a variety of sources or source types, or collaborating with their peers on more innovative technology to create websites, blogs, podcasts, etc. On their end, students sometimes take for granted what they encounter on their phones, not always understanding or examining through a critical lens how—as Safiya U. Noble and Ruha Benjamin have studied—the internet amplifies hatred, racism, and sexism (Noble, 2016). Mobile literacy, then, does not by itself help students learn. These students are only going to have the skills necessary to succeed at the college level if steps are taken to widen their perspectives on and use of technology in various forms. This all ties into another type of literacy.

Information literacy focuses on, "the ability to locate, access, evaluate, and use information that cuts across all disciplines, all learning environments, and all levels of education" (Standards, 2000). This is something we often dwell on in academia, especially as it relates to things like finding articles and books for projects, or the analysis of those resources. Reflecting on my own college experience a little over a decade ago, I don't even need a full hand to count the few teachers who took time out of their lecture schedules to walk us over to the library, to go over how to do research on the college's databases, how to use the school archives, etc. I can't emphasize enough just how important those rare building-block experiences were for me as a History major. When we talk about information literacy though, and when we talk about the importance of technology in our lives, those skills and learning opportunities are essential not just for Liberal Arts majors, but for all our students.

Regardless of our individual disciplines, we as educators need to create systems that create more equitable learning environments for our students. Stanley Aronowitz noted that literacy (writ-large) for Freire, "was not a means to prepare students for the world of subordinated labor or 'careers,' but a preparation for a self-managed life (2009, cited in Giroux, 2010)." bell hooks, building on Freire, similarly argued that education is "about the practice of freedom," creating classroom environments in the process where, "students are often presented with new paradigms and are being asked to shift their ways of thinking to consider new perspectives," especially those related to class, race, and gender (hooks, 1994). Educating students today should similarly adopt and encourage new technology frameworks and mindsets by simultaneously demonstrating to our students the opportunities in these systems, as well as the malleability and danger in these systems, and how inequitable digital opportunities can harm them and others. This sort of critical digital pedagogy, in combination with providing students with the tools they need to close preexisting digital opportunity gaps, can give students an education they can carry over into the rest of their lives. This is where a third type of literacy comes into play.

Digital literacy is about more than just the skills we want students to develop—it's the application of those skills and knowing when and how to incorporate them into people's everyday lives. Maha Bali argues that digital literacy focuses on the when, why, who, and for whom you use these skills and tools. These are the deeper, more critical questions we have to get students thinking about if we really want them to excel with digital technology and online mediums. After we know they're comfortable with those tools it's connecting to questions like, "When would you use Twitter instead of a more private forum? Why would you use it for advocacy? Who puts themselves at risk when they do so?" (Bali, 2016). The point is to go beyond the academic application of those tools and get to a point where that technology can affect change in their personal and professional lives.

To put it another way, this knowledge (not just information, but the application of that information) has to change how our students interact with the world around them (Chetty et al., 2018). It does not matter, for instance, if your university offers four different workshops on how to use a specific program like Word, Photoshop, or Excel; probably every educator has stories of offering opportunities that students didn't attend or find valuable. What matters is helping students close opportunity gaps by demonstrating how those digital services have real-world uses. Using the example of Excel—students won't care about learning how to write formulas and set-up tables if they're just elements to a core curriculum class they need to pass to graduate. What they will care about, though, is learning how to use those techniques to calculate their college loans and interest rates, or using those techniques to create a budget so they can afford to study abroad, or how to keep track of expenses and proceeds from a Greek-life fundraiser, etc. Or, maybe that tool can be used to help a student who lives in an apartment off campus—or who has a family, or who struggles with their money, etc.—make a budget and forecast their expenses over time so that they can become financially independent.

This more holistic investment in our students has to be rooted in dialogue and a desire to help our students grow based on their needs. This aligns with bell hooks's call to action (building on the words of Tich Nhat Hanh) regarding the importance of "practice in conjunction with contemplation," where education should call "on students to be active participants, to link awareness with practice" (hooks, 1994). It is only once we get to know our students that we can help them apply the tools, skills, and knowledge at our shared disposal. For those students on the margins, we can work together to buck the institutions and systems that let this digital divide grow, where they can find a whole swath of new applications for various technologies and become more independent, life-long learners. What that interaction and focus looks like will be different depending on your discipline, but there are some basic approaches we can all take.

Teaching Strategies for Closing the Digital Divide

There's a lot we can do to help our students close the digital divide. But before we can dive into metacognition, various literacies and learning exercises, low-stakes assessments, or advanced semester-long projects, we need to assess for the basics: what do students already know/ don't know how to do? In the same way that we wouldn't start a History course on World War II with the Battle of Stalingrad, or begin a Physics class with Newton's Third Law of Motion without also going over the First and Second, we also shouldn't start off our semesters by expecting students to complete assignments or act in digitally responsible ways that they don't have any context for or experience with. I'm referring, first off, to the skills that should have been (and probably were) covered during their first-year student orientation.

Anyone who has ever taught at a university can tell you about the student who never responded to their emails, inevitably leading to struggles and even conflict toward the end of the semester. Assessing whether students are accessing their email or their online class portal by having them complete beginning-of-the-semester checklists or ungraded quizzes, for instance, can establish an important baseline for the rest of the semester, helping us create better interventions and learning opportunities as they arise rather than after it's too late. These sorts of exercises are valuable both for first-year students still acclimating themselves to college or for upperclassmen who never developed the habit of checking their email. In either case it's a red flag that there may be some other gaps in their knowledge that need to be addressed.

From there we can move on to assessing their literacy and research or discipline-specific skills. Similar to the above, we can keep the emphasis here on making these ungraded or low-stakes assessments, where the focus is on improving student skills and helping them grow rather than on potentially punishing them with low grades. It could start as an open-ended question or a low-stakes homework assignment, where you ask them, "How do you determine whether a source is reliable or not?" More helpful, though, is if they complete more interactive exercises like concept maps, where they draw in the center of a paper the central idea, like "How do we understand the past?" or, if you are working toward a specific assignment, "How do you complete a research paper?"

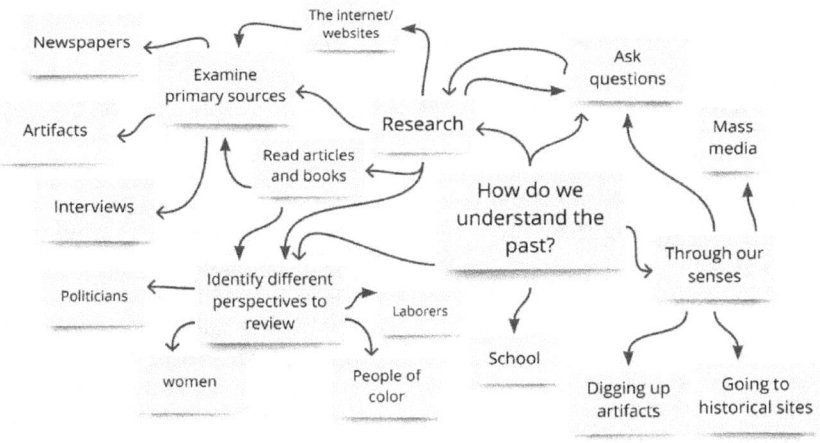

Figure 1: Concept Map from Early American History lesson following in-class exercise, Fall 2021. Made using miro.com

If you have an assignment in mind, like a research paper, from that central point they can expand out into the various elements they might need to complete the assignment, like finding and evaluating sources, taking notes, synthesizing information, writing a rough draft, etc. Given the emphasis on dialogue earlier, such an exercise can then be workshopped with the full class, where you compile their answers together on the board in a more comprehensive map. From there you can determine whether they have a sense for what you want them to do, allowing you to provide further exercises like sequence chains, where you can string those various elements of a research project into an order they can follow.

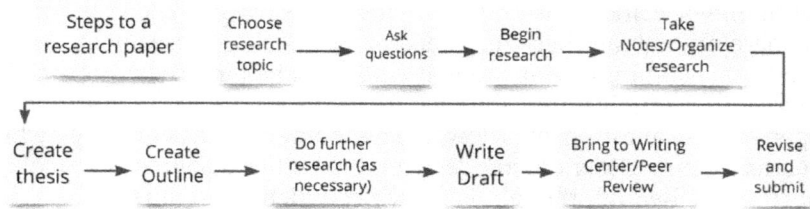

Figure 2: Sequence Chain from Early American History lesson and exercise. Made using miro.com

conversations both in and out of class. In the example of a research paper, when a student notes the need to gather sources, you can push them further (perhaps even with the help of a school research librarian)—what is an acceptable resource for a research paper? Depending on your discipline, is an article for a peer-reviewed academic journal or news organization expected? What about .edu or .gov webpages? Does recency of the piece matter?

Maybe this can stimulate a deeper conversation about the biases in specific search engines, their manipulability, the harmful ways they can collect data, and the pros and cons of using them compared to, for instance, a college library database.[5] Perhaps this leads to an exploration into how even college databases can be problematic, where restrictive acceptable use policies lead to a different form of digital redlining that disproportionately affects community college students (Gilliard & Culik, 2016). All of these topics are important entry points for getting students to think more broadly about digital literacy, accessibility, and the impact all these things have on their and others' lives.

Many Composition and First Year Seminar teachers are already accustomed to getting students to think more metacognitively while training them to do work across different genres, helping them understand the wide reach and applicability of their writing while using educational tools in new ways. Why can't every class do that, especially if what we want them to do is still academically rigorous, or mimics what professionals in our fields are doing out in the "real world," outside of just writing papers for niche audiences? As an historian I can look at how many professionals are creating blogs, engaging with audiences on Twitter, doing podcasts or vlogs, creating pop-up museums in the middle of cities or campuses—all of which can become engaging asess-

---

5. See Noble, Algorithms of Oppression, as well as Sarah Berry, "Your Complete List of 200+ SEO Ranking Factors," WebFX (June 2, 2021), https://www.webfx.com/blog/internet/seo-ranking-factors/

ments and content to dive into with my students while acclimating them to the wider uses of the technology at their fingertips.

The important part here is shifting those assignments and curriculum so that students are going beyond passive consumption of media/content. Without naming names, we can probably all think of "innovative" tools textbook publishers provide to help assess student learning, like ready-made online quizzes. These tools, though, usually just amount to extra layers of digital surveillance, where students check-off boxes in an online portal and the scores are reported to the instructor with metrics measuring accuracy, depth of comprehension, time on task, etc. While these sorts of tools may encourage a banking model of education, they do not engage with our students or help them become more independent learners.

If we are going to use technology to help our students learn, and to help them close digital opportunity gaps, we have to go about it in a way that encourages them to engage with their work and become active producers of knowledge, where they can think more critically about their work, the mediums through which they operate, and the wider contexts for their work (Bensen, 2015). This can and should tie into the higher end of Bloom's Taxonomy (and Freire's disdain for the banking method in education), of going beyond just standardized tests and assignments by instead making students assess, synthesize, and create new works that are informed by their own experiences, interests, and skills.

Take the discussion forum, a tired medium designed as much to measure online or hybrid-class attendance as it is about class participation (Morris & Stommel, 2018). The standard model asks students to post an initial post early in the week and then follow it up with one or two response posts of a shorter length later in the week. Despite our attempts to avoid vapid agreement and repetition of ideas, these forums rarely simulate the discussions that would happen in face-to-face classes. As such, they usually reinforce being rewarded for meeting the bare minimum (citing x-number of sources, writing at least z-number of words, etc.).

We can and should reshape online discussions by encouraging students to stretch their creative thinking muscles and by designing these assignments to mirror online and social media forums they likely interact with on a daily basis—though in a more thoughtful way. One idea I experimented with at the outset of the pandemic was having students finish each of their initial discussion posts with a question to their online classmates. The question could relate to the specific sources I asked them to read, the topics of the week, or even bigger-picture questions that analyzed course themes. As part of their response post each student was required to address a peer's question. The production of and answering of these questions generated more critical thinking and reference back to that week's primary and secondary sources (and lecture videos) than what I am used to seeing. These question-and-answer sections also provided me with more opportunities to address confusion around specific topics, to connect to the works of other historians, and to make the forums feel more like a place where ideas were exchanged rather than just act as a chance for me to grade attendance.

Another strategy I've incorporated to make this assignment more interactive has been to ask students, as part of their initial response, to connect some of that week's readings to a relevant current event. Part of this requirement asks that they find a reliable news source and link to it in our online forums. In an anthropology class that could involve an article examining the effects of climate change on coastal communities. In a history class it can take the shape of connecting debates during the Constitutional Convention to debates today around the size and scope of the federal government. In a politics or sociology class the options are probably limitless given how tuned-in those courses are to current events. The point is to regularly assess and build on students' online information literacy skills, making them find, evaluate, and relay reliable information while connecting to that week's lesson in insightful and engaging ways.

The table below includes just a few other interchangeable examples to show the sorts of learner-centered digital and hybrid assignments that can stimulate creativity, critical thinking, and life-long digital literacy skills while still meeting the sorts of learning goals and outcomes our universities stress. Each of these provides opportunities for student choice, encouraging them to pick topics that drive their interest and perhaps the paths they want to pursue after college. If properly emphasized, our digital and hybrid assignments can provide students with an appreciation for a variety of new skills and outlooks, where the internet, its forums, tools, archives, etc., all become something useful and potentially life-changing, rather than act as "trillions of lines of code" no better than "flotsam" scattered across the web (Morris, 2018a).

And to be sure, this responsibility doesn't need to fall just on the teacher's shoulders alone. Multiple studies exist that emphasize the role of librarians in digital and information literacy training. That kind of partnership can help reduce some of the sage-on-a-stage mentality

Putting Theory into Practice	Science/Math	History	English	Anthropology	Sociology	Political Science
Assignment	Create a textbook with experiments for middle/high schoolers	Record a song about historical figures (as a satire, a rap battle, a ballad, etc.)	Create a blog for your semester's worth of work	Develop a podcast episode on a course topic	Create an online questionnaire for a sociological study	Establish a digital archive/exhibit of presidential tweets on a topic
Digital Skills Developed	Digital design/editing/publishing	Record/edit video/music on popular online forums	Web design/editing	Incorporate audio from interviews, edit into a presentation	Web design, online promotion techniques, data analysis	Social media analysis, data retrieval, web design
Course Objectives Reached	Create/solve average to complex scientific/mathematic problems	Demonstrate understanding of competing historical viewpoints	Curate semester's work into a portfolio, demonstrating semester-long writing development	Incorporate ethnographic fieldwork into a presentation	Create research study accessible to a large test population	Identify/compile government information on perspectives and topics

Table 1: Examples of assignments that can stimulate creativity, critical thinking, and life-long digital literacy skills

in our classrooms. The research points to the notion that if students receive information and digital literacy training in a more collaborative manner, and if they are receiving those lessons from different parts of your institution, they might benefit more from those opportunities while bringing them all up to a similar baseline (Bensen, 2015).

Likewise, IT department trainings are another wonderful set of opportunities for collaboration. Having attended many of them over the years I know firsthand how thorough and attentive those staff members can be when it comes to helping our campuses better understand the tools at our disposal. As teachers we can support those endeavors with more than lip service by including those opportunities into our own lessons, by having guest speakers, by offering our own how-to lessons and materials, or even by providing class waivers if a student needs to attend a workshop. The infrastructure to support our students is already at our finger-tips—we just need to put it to use.

Conclusion

We don't need students to rely on Microsoft Word to create scholarly projects; we don't need to live and die by PowerPoint presentations; we don't have to settle for some of our students graduating with the same general skills (or lack thereof) they entered our classes with. As Ruha Benjamin suggests, we need to learn from the experiences of the Coronavirus pandemic's effect on education, and that "without a deep engagement with critical digital pedagogy, as individuals and institutions, we will almost certainly drag outmoded ways of thinking and doing things with us" (Benjamin, 2020). By expanding how we think about digital skills and classrooms, and by providing students with the opportunities to nurture skills and mindsets they can use for the rest of their lives, we can do a tremendous amount of good in at least starting to close digital opportunity gaps and the digital divide.

Acknowledgements

I would like to thank this anthology's editors, as well as Raymond Dinsmore, Dr. Lila Teeters, and especially Ian S. Wilson for very timely and helpful suggestions at various stages of this work. This piece would not be what it is without their help.

References

American Library Association (2000). ACRL standards: Information literacy competency standards for higher education. *College & Research Libraries News*, 61(3), 207-215. https://crln.acrl.org/index.php/crlnews/article/view/19242/22395

Associated Press (2011, February 8). Smart phones outsell PCs for the first time. *CBS News*. https://www.cbsnews.com/news/smart-phones-outsell-pcs-for-the-first-time-7330519/

Associated Press (2019). Homework gap' shows millions of students lack home internet. *NBS News*. https://www.nbcnews.com/news/us-news/homework-gap-shows-millions-students-lack-home-internet-n1015716

Bali, M. (2016). Digital skills and digital literacy: Knowing the difference and teaching both. *Literacy Today*, 4(33), 25.

Benjamin, R. (2020). Foreword. In J. Stommel, C. Friend & S. M. Morris (Eds.), *Critical digital pedagogy: A collection*.

Bensen, E. (2015). *Multimodal composing across disciplines: Examining community college professors' perceptions of twenty-first century literacy practices* [Doctoral Dissertation, Old Dominion University]. ODU Digital Commons. https://digitalcommons.odu.edu/english_etds/5/

Braverman, B. (2016). The Digital Divide: How income inequality is affecting literacy instruction, and what all educators can do to help close the gap. *Literacy Today*, 33(4).

Chetty, K., Qigui, L., Gcora, N., Josie, J., Wenwei, L. & Fang, C. (2018). Bridging the digital divide: measuring digital literacy. *Economics*, 12(1), 20180023. https://doi.org/10.5018/economics-ejournal.ja.2018-23

Cortez, M. B. (2017). 21st-Century classroom technology use is on the rise. *EdTech Magazine*. https://edtechmagazine.com/k12/article/2017/09/classroom-tech-use-rise-infographic

Dixon-RomÁN, E. J., Everson, H. T. & Mcardle, J. J. (2013). Race, poverty and SAT scores: Modeling the influences of family income on black and white high school students' SAT performance. *Teachers College Record: The Voice of Scholarship in Education*, 115(4), 1–33. https://doi.org/10.1177/016146811311500406

EdTech Staff. (2021). The checklist: Bridging the digital literacy gap. *EdTech Magazine*. https://edtechmagazine.com/higher/article/2021/03/checklist-bridging-digital-literacy-gap

Eubanks, V. (2011). *Digital dead end: Fighting for social justice in the information age*. The MIT Press.

Freire, P. (1970). *Pedagogy of the oppressed* (M. B. Ramos, Trans.). Continuum.

Gilliard, C. & Culik, H. (2016). Filtering content is often done with good intent, but filtering can also create equity and privacy issues. *Common Sense Education*. https://www.commonsense.org/education/articles/digital-redlining-access-and-privacy

Giroux, H. (2010). Rethinking education as the practice of freedom, Paulo Freire and the promise of critical pedagogy. *Policy Futures in Education*, 8(6), 715-721. http://journals.sagepub.com/doi/pdf/10.2304/pfie.2010.8.6.715

Hockly, N., Dudeney, G. & Pegrum, M. (2014). *Digital literacies*. Routledge.

hooks, bell. (1994). *Teaching to transgress: Education as the practice of freedom*. Routledge.

Lai, J. & Widmar, N. O. (2021). Revisiting the digital divide in the COVID-19 era. *Applied Economic Perspectives and Policy*, 43(1), 458–464. https://doi.org/10.1002/aepp.13104

Mooney, T. (2018). Why we say "opportunity gap" instead of "achievement gap." *Teach for America*. https://www.teachforamerica.org/one-day/top-issues/why-we-say-opportunity-gap-instead-of-achievement-gap

Morris, Sean Michael. (2018a). Courses, composition, hybridity. In Sean Michael Morris & J. Stommel (Eds.), *An urgency of teachers*. Hybrid Pedagogy.

Morris, Sean Michael. (2018b). Critical instructional design. In Sean Michael Morris & J. Stommel (Eds.), *An urgency of teachers*. Hybrid Pedagogy.

Morris, Sean Michael & Stommel, J. (2018). The discussion forum is dead; Long live the discussion forum. In Sean Michael Morris & J. Stommel (Eds.), *An urgency of teachers*. Hybrid Pedagogy.

Noble, S. U. (2016). A future for Intersectional black feminist technology studies. *The Scholar & Feminist Online*, 13.3-14.1. https://sfonline.barnard.edu/traversing-technologies/safiya-umoja-noble-a-future-for-intersectional-black-feminist-technology-studies/0/

Pendakur, S. L. (2016). Empowerment agents: Developing staff and faculty to support students at the margins. In S. R. Harper & V. Pendakur (Eds.), *Closing the Opportunity Gap: Identity-Conscious Strategies for Retention and Student Success*. Stylus Publishing.

Rich, M., Cox, A. & Bloch, M. (2016, April 29). Money, race and success: How your school district compares. *The New York Times*. https://www.nytimes.com/interactive/2016/04/29/upshot/money-race-and-success-how-your-school-district-compares.html

Richars, E., Aspergen, E. & Mansfield, and E. (2021, February 4). A year into the pandemic, thousands of students still can't get reliable WiFi for school. The digital divide remains worse than ever. *USA Today*. https://www.usatoday.com/story/news/education/2021/02/04/covid-online-school-broadband-internet-laptops/3930744001/

Selyukh, A. (2016, February 6). How limited internet access can subtract from kids' education. *NPR*. https://www.npr.org/sections/alltechconsidered/2016/02/06/465587073/how-limited-internet-access-can-subtract-from-kids-education

Skinner, Benjamin, Hazel Levy, and Taylor Burtch. (2021). Digital redlining: the relevance of 20th century housing policy to 21st century broadband access and education. (EdWorkingPaper: 21-471). https://doi.org/10.26300/q9av-9c93

Stanton, L. (2017). NDIA, CYC accuse AT&T of "digital redlining" in low-income cleveland neighborhoods (Telecommunications Reports).

Zumbrun, J. (2014, October 7). SAT scores and income inequality: how wealthier kids rank higher. *The Wall Street Journal*.

Sharing Instructional Design
Collaboration and community with the past, present, and future

Mary Klann, Logan Gorkov, and Rossel-Joyce Garcia

We collaborated on the writing of this piece, but have also maintained our own distinct voices in the notes. Throughout the piece, readers will find footnotes labeled with our names to indicate authorship. These notes are a nod to the social annotation that was a large part of the course we'll discuss in detail in the article below.

Mary Klann: I am an adjunct professor of history at UC San Diego, San Diego Miramar College, and Cuyamaca College. I received my PhD in US History from UCSD in 2017. My research and teaching interests include Native American History, Women's History, and Digital History. I love teaching, especially teaching online.

Logan Gorkov: I graduated from UC San Diego in Spring of 2021 with a Bachelor of Arts in History and Studio Arts. My research interests include Indigenous History, LGBTQ+ History, and decolonization movements. I have a love-hate relationship with academia, but I do believe that everyone should have access to and control over their own education.

Rossel-Joyce Garcia: I am a class of 2021 graduate from UC San Diego, where I received degrees in Ethnic Studies (B.A) and History (B.A.) My research interest is K-12 education in the United States, focusing primarily on the entanglements of neoliberalism, race, ethnicity, and class in education.

Introduction

In the article that follows, we make an argument for sharing the responsibility of instructional design with all members of the classroom. Our case study is our class, Digital History and Memory, (nicknamed "DigHist") which took place at University of California, San Diego from January 2021–March 2021. Throughout the course we learned about the power of collaboration and community building between all members of the classroom in shaping course design and outcomes.

Our DigHist course set out to explore the relationship between digital technology and historical research, writing, and memory. During the first four weeks of the 10-week quarter, we collaboratively analyzed existing digital history projects, especially those created by university students, such as the Race and Oral History Project at UCSD.[1] For the next six weeks, smaller groups of students analyzed selections from a digitized collection of 21 different student newspapers (a total of 675 digitized items) available from the UCSD Library Digital Collections. We were lucky to receive an introduction to the collection from Heather Smedberg of the UCSD Special Collections and Archives.[2] We ended the course with a collaborative proposal for a future digital history project featuring the student newspaper collection.

Our essay below will explore a few important things about the class structure and modality. There were 32 student participants, and one instructor. Due to the COVID-19 pandemic, our class was conducted online and asynchronously. We did not have synchronous meetings, communicating instead using tools for social annotation (Hypothesis), collaborative discussion (Padlet, Google Docs), and peer feedback (sharing in progress work via Padlet and a class WordPress blog). The class assignments were also ungraded. For our course, ungrading was characterized by a combination of instructor and peer feedback and students' "declarations" of their finished work. After finishing annotations, assignments, and/or providing feedback to peers, students submitted their declarations to Canvas, which automatically calculated their point totals.[3] Instructor feedback was still given for assignments, and academic integrity regarding the declarations matching up with the

---

1. The Race and Oral History Project is the result of the oral history research conducted by members of UCSD's Race and Oral History in San Diego course. Students in the course worked with community partners, organizations devoted to themes such as migration, settler colonialism, and militarism to document the history of race in San Diego.

2. **Mary Klann (MK):** Heather Smedberg, Harold Colson, and Cristela Garcia-Spitz at the UCSD Library provided essential guidance and support before the class started and as the course progressed.

3. **MK**: Credit for the language and format for "declarations" goes to Laura Gibbs, former online educator at University of Oklahoma, who has generously shared many ideas with me this past year.

amount of work completed was considered, but the instructor trusted students to make decisions about how they engaged with the course.[4,5,6]

Looking to the Past, Looking to the Future

At the end of our DigHist course in March 2021, students wrote brief letters to future students with advice on how to approach the course. Many of the letters touched upon the value of creativity, exploration, risk-taking, and freedom in the course, with specific mentions of the ungrading policy, collaborative learning environment, and lack of fixed deadlines. One comment highlighted a particular aspect of the course's design: "Take advantage of the 'structurelessness' of the class and learn all that you can!" The class wasn't unstructured in the sense that students had complete autonomy over the subject of the digital history project (the sources we used were part of a specific digitized collection), but in many important ways it was "structureless."[7] Primarily, we didn't set out to end with a completed digital resource. We would end with a proposal that could be taken up, edited and revised, even rejected, by future students. Looking back, the open-ended "work-in-

---

4. **Rossel-Joyce Garcia (RJG)**: This was the first ungraded class I have experienced. Despite students not being graded, the participation levels in this class were similar, if not greater than, participation levels in the other classes I've taken at UCSD.

5. **Logan Gorkov (LG)**: Other professors I had encountered online during the pandemic have also taken a more "hands off" approach, where there were little or no synchronous lectures. However, one such professor gave no instruction novel to the course, he instead used pre-pandemic recordings of the class, where discussion between the professor and students could not be heard (apart from the professor's contextless answer). Feedback given on assignments was minimal—I received a total of 7 words between the midterm and final papers, and that is the extent of my interaction with that professor.

6. **MK:** I started ungrading all of the courses I teach (at UCSD and other colleges) in Fall 2020. The quality of students' work and effort has been the same, if not better, than courses I have taught with traditional grading methods. I'm never going back to traditional grading

7. **MK:** Before the course began, I envisioned ending our quarter with a finished digital project. It was only after talking to the archivists and librarians at UCSD that I fully understood the magnitude of all that would entail. I decided to adopt a more open-ended approach to the "end" of the course. Our final project was defined further as the quarter progressed, and we began to identify key themes as a group. The class contributed to a shared Google document where everyone contributed answers to questions that had been generated over the course of the quarter. I am teaching the course again in Spring 2022, and the new group of students and I will build off this Google document.

progress" design of the course might have helped to establish unexpected layers of community.

The first layer of community was our class, the 33 participants, including the instructor, who shared reflections and analysis of existing digital history resources, examined the primary sources, and built the final proposal. We will go in-depth on this topic in later sections of this article, but it is worth mentioning that this was not a typical classroom where private work was submitted by students for private feedback from the instructor. Every aspect of the course was open to collaboration, and the weight of instructor feedback was not necessarily greater than that of fellow students. Everyone brought a unique perspective based on the individual research done, and connections were made organically between different students' work and how it might fit into the bigger picture.

The second layer was the greater UCSD campus history and community, as the primary sources, student newspapers, prompted many participants to look critically at our own experiences on campus, especially in the context of the COVID-19 pandemic and learning together in digital space. The community we formed was with each other, but also with the students of the past, who shared many of the same traditions, protests and fights with administration, global awareness, and desire to change the campus for the better. In their final reflection, one student wrote, "I learned so much about UCSD's past from reading through the newspapers, and I had no idea that UCSD had so many student publications and editorials in the past!" The same student emphasized the value of examining digital history (which, as our class found, could mean anything from the methods used to store and catalog the sources to the ways that visitors interacted with those sources), especially because "we are living in the age of technology." COVID-19 only made our experiences of the "age of technology" more acute.[8]

The last layer was more future-oriented. Since we designed something that was meant to be implemented, revised, and edited by future students, we were communicating with groups who we hadn't yet met, who would be experiencing the course in a different political and social moment in time. Additionally, many students asked for ways to follow

---

8. **MK**: I had already been trained to teach online pre-COVID, through training provided by the San Diego Community College District. I love teaching online. However, I knew that many students in the course had not taken online courses outside of remote learning during COVID and did not have the same feelings about online learning. I found all the learners in this class to be incredibly willing to critique and engage with digital technology and how it shapes historical research.

up with the project, to engage in the future as a participant. Many of us looked towards the future (either as instructor, student, or alumni) with a desire to see how the project progressed.[9]

Ungrading: An Opportunity for Growth and Discovering Genuine Interest

Ungrading in this course was significant to the experience.[10] Ungrading especially contributed to community building, because there was no competition involved. Students collaborated without the worry of a group project where the grade depends on if everyone does their fair share, but also without the pressure of having to do the entire project right on the first go. Students were able to take risks and still gain valuable feedback, rather than get the grade but gain very little growth.[11] In their letters to future students, one student connected ungrading to the

---

9. **LG**: This was the first time I was interested in continuing to hear about a project after my time in a particular class ended. If I hadn't actively participated in classes, I would know the blame for that was on me, but this is the first and only time during my academic career where I felt like I was doing something that made an impact on a future other than my own!

10. **MK**: I owe my understanding of ungrading to several educators who have written about their methodologies, especially (but not limited to): Gibbs, L. (March 15, 2019). *Getting rid of grades.* OU Digital Teaching. http://oudigitools.blogspot.com/2019/03/getting-rid-of-grades-book-chapter.html; Kohn, A. (2018). *Punished by rewards: The trouble with gold stars, incentive plans, A's, praise, and other bribes* (25th Anniversary Edition). Mariner Book.; Sackstein, S. (2020). Shifting the grading mindset. In S. Blum (Ed.) *Ungrading: Why rating students undermines learning (and what to do instead)* (pp. 74-81). West Virginia University Press.; and Stommel, J. (March 11, 2018). *How to ungrade.* https://www.jessestommel.com/how-to-ungrade/.

11. **LG**: In college, I personally became so obsessed with grades that if finals were available to be picked up after grading I never went out of my way to do it. The grade I got in the class was all that mattered, and I could already see that online. Improvement didn't matter—and it did not feel like improvement was valued between courses regardless. Every professor has such distinct expectations that frequently do not even line up with what is taught in the required core writing courses. A paper was not about becoming an active participant in my learning and research, but to say what a professor wanted to hear. That isn't to say I never looked at feedback—but the majority of feedback I received from professors (or their graders) was so baffling that I didn't know what to take from it except for how to write my next paper for that same person. I truly feel like I built on a unique writing skill here, not because I was writing to please Mary Klann, and not because I was writing another discussion board post, but because of the novel method of the blog posts and genuine care into this project—and us as students, having potential to Make (with a capital "M") something that mattered, not just demonstrate what we have learned, or rather, how we have learned to appeal to a professor.

creative process: "you can take risks knowing that you will be academically safe even if you later decide it wasn't the right direction!"[12] The safe route was not necessary, and in this class where the students depended on each other as much or more than they did on the instructor, there was always available feedback in the path to a project proposal rather than the judgment of worth that comes with a grade.

Another aspect of ungrading in the success of our project proposal—and hopefully future project[13]—was the element of being able to pursue information based on the group's interests. We worked from a set primary source base, the student newspapers. But, the class chose which specific issues to focus on within that set of primary sources. It is unlikely we would have seen the same impact if the overarching project theme had been unilaterally "assigned" by the professor. The project itself remains open-ended for future groups of students depending on the sources that interest them the most. In this sense, ungrading represented freedom for each student to fully devote their research time to elements of the sources that interested them. The letters to future students reflect this. One student wrote that ungrading "made doing the work of reading and engaging with [the student newspapers] much more enjoyable and I was able to put them into conversation with things I already knew and had concurrently been learning in other classes." Students were drawn to the sources they had connections to, where they felt they had something important to add, and that made the final proposal all the more meaningful and compelling.

Understanding Digital Literacy/History in the Context of COVID-19

As we reflected on the course and began to write this piece, a question emerged about the context of COVID and its impact on our instruc-

---

12. **RJG**: One of my biggest takeaways from the DigHist class was finding what my non-essay writing voice sounded like, and one of the primary reasons I was able to do that is because of the ungrading policy. I had the freedom to play around with my writing without worrying about what would happen to my grade! In the end, I had written some things that I didn't really like, but I also wrote some things that I absolutely loved, and in the process, I learned a lot about my writing style.

13. **MK**: Our class was listed as a "special topics" class in US History. As an adjunct, my employment at UCSD and the other colleges where I teach is precarious. However, I hope to someday offer this course as a permanent course with its own course number, so future students (and alumni) may participate in this project. My goal is to continue to offer the course (at least partially) online.

tional design: Is this a class that could have been taught the way it was or been as successful as it was if it weren't for the approximately 9 months of practice we already had learning online?

This question has a different answer depending on each individual participant's perspective. One of the first questions we considered as a class was what was digital literacy? If we are able to navigate the internet with ease, are we digitally literate? Some students asserted that they were, as one annotated in the margins of the American Historical Association's definition of digital literacy (American Historical Association, n.d.): "I definitely would consider myself digitally literate, especially in the last year when practically my whole life has been based on online meetings, assignments, and maintaining friendships remotely." However, others expressed fear and trepidation around the idea of "digital literacy," and still others acknowledged there were some areas of the internet and certain tools that they continued to learn more about. Multiple students also noted that they might assume a level of digital literacy because "a lot of us grew up using technology everyday," and many "often assume since they grew up with technology, that they know absolutely everything about it."

Regardless of where students fell on the scale of digital literacy, the reality was that most, if not all, students surely had some experiences with technology over the course of the COVID-19 pandemic, as classes shifted to remote learning. As a result, there seemed to be more of an organic use of the digital space in this class compared to a pre-pandemic in-person class that used the same digital tools.[14] In the latter, students may have never had to worry much about digital literacy and may also have had ample opportunity to learn and engage with class material outside of digital spaces, such as in the physical classroom. However, in this class, students were required to step into these digital spaces alone, just as they had been required to in the previous nine months. Since this class was only offered online, unlike a class offered in-person, there were no alternative spaces of learning. If students desired to partake in the class, whether it be because they needed the class for a general education requirement or because they genuinely wanted to learn about digital history, there was really no other option

---

14. **RJG**: Rossel-Joyce Garcia: In the majority of classes I took pre-pandemic, the primary digital platform used was Canvas. One of the most common uses of the platform was for students to create original discussion posts and respond to other students' posts. In my experience, this use of the platform has often resulted in bare-minimum discussion posts and responses that felt forced. Even if students got little to nothing out of it, they were able to depend on future in-person classes to interact with the material.

Designing for Care

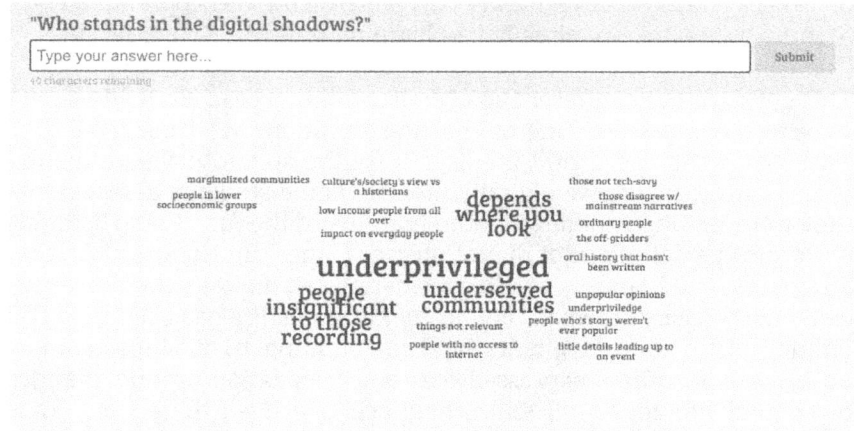

Figure 1: Student-generated word cloud in response to the question "Who stands in the digital shadows?"

but to step into the online "classroom."[15] The product of these circumstances was a class of students that seemed to embrace (some fully willing, some reluctant, and some in-between) the digital nature of the class.[16,17]

However, as we worked out during the first few weeks of the course, access to digital literacy and engagement with digitized archives of historical material is contingent on the social, cultural, racial, and/or

---

15. **MK**: As an instructor, this tension is something I really tried to keep in mind as the course progressed. My goal with this course (and with all the courses I teach) is to try to cultivate a space for joy in learning, whatever that may look like for each student.

16. **RJG**: I was one of the students who was quite willing to embrace this! One of the things I missed most when switching to online learning is the in-person discussions that were common in many history classes at UCSD. Since there was no way for me to get this face-to-face interaction in class during the pandemic, especially because our class was completely asynchronous, I was eager to accept the digital "classroom" and whatever opportunities it had for interaction with my classmates.

17. **LG**: I was in-between initially. I had taken face-to-face classes with Professor Klann before, and so I was familiar with her digital tools, but hadn't really experienced them as a primary method of communicating with others. Despite having been in 9 months of online classes already, I had yet to experience a sense of community in my classes (even those that did have live meetings), and so I came in skeptical. However, by the time we got to working on the student newspapers, I was excited to see feedback from my classmates in a way that I hadn't experienced before.

economic spaces we occupy. In her 2016 article in the *American Historical Review*, historian Lara Putnam, questioning the tendency of more and more historians to engage with digitized primary sources, wrote, "There is a real world out there. The totality of sentences that have ended up in print in no way corresponds to the proportions of past human life. Who stands in the digital shadows (Putnam, 2016)?" The class generated a revealing word cloud[18] in response to Putnam's question, highlighting the role of the historian/researcher in framing what is relevant and important, and the ways that class, race, and/or gender identity impact how historical actors' experiences are remembered and recorded.

As a whole, the annotations of our readings in the first three weeks of the course revealed our willingness to critically engage with all the contradictions of digital history, questioning how digital sources and tools shaped our learning, and the processes of preserving and analyzing diverse histories. In an opinion-based poll we answered during our second week, the class was given a choice of two options.[19] Did digital history have more potential to expose and document voices from underserved and underheard communities? Or, did it have more potential to leave certain historical actors in the "digital shadows"? The results were revealing: 80% of respondents voted for the first option, that the benefits outweighed the potential pitfalls of digital history. However, many people were torn. One student elaborated, "I think I am more pessimistic when it comes to digital history just because in the last year, I have seen so much of the news focused on the wrong aspects of events, ignoring the bigger picture that would have resulted from genuinely knowing the opinions of those in the 'digital shadows.'" Issues of equal internet access and other lived experiences made this question especially thorny. Other students explained that "If the community is criminally under-heard there's a lot of good-intentioned people willing to use the internet to help make their voices heard;" and "many people have found community through the internet due to not being able to find it where they live in real life (whether it be due to location, lack of representation, etc.)."

The question of the role and potential of digital history for exposing and documenting underheard communities is even further complicated by the ways in which social science research has often been violent and extractive in these communities. Western academia has long upheld a

---

18. For this class, word cloud questions were administered via Answer Garden, a free word cloud generator. https://answergarden.ch/.

19. In this class, opinion-based poll questions were administered via Slido. https://www.sli.do/.

dichotomy of researcher and subject, where the researcher is the primary producer of knowledge about the subject and the primary benefactor of the produced knowledge. In this understanding, the researcher holds the power—and the institutional means—to decide what will or will not be said about the subject. The possibility of this dichotomy certainly remains in the realm of digital history (and it is up to us as researchers to wrestle with what it means to be an ethical researcher), but at the same time, the possibility to depart from it arises. With the internet, and the practice of digital history, it is possible for ordinary people[20] to assume the role of knowledge producer through creating and sharing digital content. In this model, ordinary people, who may have otherwise been the subject about which knowledge was produced, decide how they want to be known without the same level of institutional barriers they might come across at, for example, the university. However, this model assumes that all ordinary people have access to the digital world, which is not the reality. Even if people do have access to the digital world, the question might then shift to who will be listened to. Websites like Wikipedia where anyone can add or remove information have long been discredited by academics.[21,22] What does it mean for these kinds of websites to be discredited when these websites are where ordinary people go to share digital knowledge (Rosenzweig, 2006)? This concept can even be applied to our digital history class, in which one participant was a professor of the university and the other 32 were all students enrolled in the university. Is our project worth looking at simply because it was constructed and executed within the parameters of UCSD? Would it still be worth looking at if it was done outside of the university or outside of any other research institution?[23]

It was worth digging into these complicated issues before we started building our own digital history resource. Without understanding the methodological and practical complexities of digital history, we

---

20. "Ordinary people" is used to refer to people outside of research institutions.

21. **RJG**: I remember until my sophomore year of college, teachers have told me to avoid using websites that end in .com and .org when sorting through the internet. I was always told to use Google Scholar or websites that ended in .edu, as they were thought to be more credible.

22. **LG**: There are so many instructors out there that are still afraid of internet resources (like Wikipedia) rather than teaching us how to critically engage with them. The conversation never seems to go beyond "not everything on the internet is true."

23. **LG**: Here's my personal struggle with academia—I want to do research, and I want to be "qualified" to do research, but I don't see academia as a place to do that without being extremely, impossibly selective about where I go and what structures I buy into.

wouldn't have been able to fully engage with the digitized primary sources we worked with in the second half of the class. Discussing some of the theoretical questions behind digital history helped us to understand what was behind the choices we made in our final proposal. It also helped to critically engage with other aspects of our lives that we lived online, which, during this time, was practically all of it. Moreover, discussing issues of access to digital archives and whose voices are remembered and preserved prompted us to think more critically about access to digital technology in the online classroom, and how we envisioned making our proposed digital history resource accessible.

Building a Humanized Online Classroom

In their letter to future DigHist students, one student wrote:

> "I think one thing that is encouraged a lot in this class is collaboration. We have to read over each other's work and posts all the time. Personally I think this is pretty helpful, it lets me see what's interesting to other people as well as what the common themes in what we're studying are. I think the purpose of digital history is all about communicating experiences and viewpoints."

The elements of instructional design that this student highlighted—collaborative reading, peer feedback, working together to identify common themes—corresponded well with their view of the subject matter of the course, digital history. During the first few weeks of the course, we analyzed and critiqued a selection of existing digital history projects, interrogating digital history methods such as tagging and non-linear pathways of reading. As a group, we collaboratively annotated different digital history sites using Hypothesis. Additionally, the instructor posed questions in the annotations to be answered via poll or word cloud. Students then checked in on a weekly Padlet where they could see the results of the polls and word clouds and offer their thoughts on the conversation as it evolved. As we discussed digital methods together, we established the basic goals for our own digital history resource: collaboration, frequent feedback, and consideration of the widest possible audience.

The course was heavily collaborative, but also prompted each individual to independently engage with the material, drawing out their own research interests. In addition to social annotation, students each conducted individual research on a set of student newspapers and shared that research through in-progress "project journals," documents shared through Padlet, and as shorter posts on the course's WordPress blog. As another student wrote in their letter:

> "Focus on what you find interesting when doing research. Be selective with your time and what you want to look for. There is a lot of information that you are going to have to look through and it makes it so much easier to be selective on what you want to look at. This class does a good job about giving you that freedom, but it also forces you to choose. Be selective; pick the things that are interesting to you."

Providing opportunities for individual choice and autonomy were critical, not antithetical, to the success of a heavily collaborative class. An excerpt from another letter to future students explains why: "You will be surprised your other classmates will be responsive and helpful! Be organized, provide pictures, and show your thoughts from your writing! Let them hear your voice!"

This student, as well as many other students in the class, boldly voiced their thoughts through their own work, but also in response to their classmates. One reason for this may have been that the considerably flexible timetable of the class (with the primary restriction being the quarter system) allowed students the time to process and edit their thoughts at their own pace. Once students formed their thoughts, they could share them with the class when they were ready as opposed to having to read something, immediately respond on a discussion board, and then continue the discussion in class the next day, all before they had a chance to fully absorb the material. Another reason students may have so willingly shared their thoughts is simply because they had the space to. Unlike a class taught in-person where they might have to compete to speak with a more outspoken student or where they might be sitting in the back of a 300-student lecture hall (out of reach from the professor), our online "classroom" offered students virtually unlimited space to share their thoughts freely, as they pleased.[24]

---

24. **MK**: To me, peer feedback was an essential part of the course. One of the ways I tried to communicate its significance was by giving ample time for feedback. Students usually had a full week to read and respond to their classmates' blog posts and project journals, without any extra reading or activities attached to that week. Additionally, the class also contributed to two "Thematic Reflection Padlets" throughout the quarter. These were shared spaces where the group worked together to identify connections and cohesive elements in everyone's individual research. As the quarter progressed, the themes of the digital project began to emerge in the connections that participants were making with each others' work.

Something very specific to the internet may also have encouraged people to express themselves: the sense of anonymity. Though we tend to think of online anonymity as a bad thing, as something that enables some to discriminate against others in public without repercussion, anonymity provided a sense of safety in our course. As mentioned, there was no fear of having to compete with the more outspoken student, because everyone began on a relatively level playing field in digital space. We were, of course, not completely anonymous.[25] We knew each other's full names, which isn't nearly as likely in an in-person class, but there is something to be said about posting something to the discussion board while alone in your own space versus feeling the 300 pairs of eyes turn towards you answering a question in a large lecture hall. We maintained the professionalism expected in the classroom, but the feedback went far beyond the expectations of "post once, reply twice" discussion methods.[26,27]

Would this course design work as well in a face-to-face classroom? Is online pedagogy critical to this course's design? In order to fully answer this question, we need to return to Putnam's conception of the "digital shadows," and our critical engagement with the idea of how digitally "literate" each class participant is. In many ways, this question can't be answered without fully understanding the context of COVID-19 and each participant's (whether student or instructor) experience with remote teaching and learning.

Issues of access during the pandemic became very prevalent—and al-

---

25. **MK**: There were opportunities for students to be anonymous in certain aspects of the class, namely in the annotations of course readings via Hypothesis. We used Hypothesis outside of our Learning Management System (Canvas), which meant that students could choose their own usernames. Some chose pseudonyms for this format. However, in other aspects of the course, like the WordPress blog, our full names were visible. Our WordPress blog was available through the university, so each participant's username was linked to their UCSD account.

26. **LG**: This is where the merit of a "post once, reply twice" seems relevant, if only it wasn't meant to be something that just proved you were alive and paying just enough attention to your online course. But our Padlets asked us to reply to other people and did not necessarily put a limit on it, but I feel like we got good engagement regardless!

27. **RJG**: I think a huge part of this was also that Dr. Klann gave us flexibility in our posts and responses. A big thing for me was that there were no word counts imposed on us, so I was able to focus on the quality of my posts and responses (which sometimes still ended up being quite lengthy). I also loved that we were allowed to respond in GIFs and emojis because it seemed to make things feel a little more personal, like in-person classes, and not robotic.

though things like laptops could be supplied temporarily by the school, a stable internet connection and a nearby timezone could not. Having no live portions to this class made it much more possible for everyone to fully participate, rather than having recordings of discussions that could be watched, but could hardly be added to in a meaningful way. For the full 10 weeks, students had access to every piece of work (not including declarations and personal reflections) created by other students. A student could post at any hour, and their voice would still matter and be a part of the discussion—there was never a point at which the discussions were "over" until the quarter was over.[28]

It is essential to consider how and why relationships are built and sustained, regardless of the modality. Some elements of the course design—for example, an unfinished final product, and ungrading—could be applied to either a fully online or fully in-person course, or something in between. For our class, which occurred in the context of COVID-19, the asynchronous online modality facilitated essential elements that helped make the class successful: collaboration, feedback, and individual agency.

A critical question we asked was how did we build a cohesive project without ever having a face-to-face or live online meeting? Many of the points already brought up speak to this, such as the space and time to boldly voice thoughts. Although students worked individually on assignments, they also had time to read and respond to what others had written. Students built on the feedback given by other students and the instructor, and the feedback they provided to others. A strong sense of communal tone was acquired, whether we were aware of it or not, and then continued to build. There was also time to absorb the feedback received and incorporate it into the next assignment, where the process started again, establishing trust in the students that is not typical for project-based classes.[29] Students could depend on each other, rather

---

28. **MK**: And here we are continuing these conversations after the quarter's end!

29. **MK**: I saw this trust build as the instructor. I also tried my best to communicate my sense of trust in everyone by sharing my feedback about in-progress work openly. Since the course was ungraded, my feedback was never meant to justify why I had subtracted points from a particular assignment. I always tried to frame my feedback around things that stood out to me as exciting or useful for the class discussion and questions that could encourage further thinking about a particular topic. I also left feedback in "public," responding to annotations, posting to the Padlet, posting comments to the WordPress blog, or creating my own public Padlets or Google Jamboards with my observations about themes that had come up for particular assignments.

than immediately going to the professor when they were feeling stuck.[30]

Building Community through Historical Student Newspapers

We never met face-to-face and had only one optional online meeting in our course. However, there was a valuable sense of community that grew over the course of the ten-week quarter. One of the ways that we built that community was through collaborative analysis of the digitized collection of historical UCSD student newspapers. Students kept "project journals" reflecting on a set of chosen newspapers, which they turned into the class twice during the quarter. Students also posted to our WordPress blog with essays about specific themes and issues that arose from the newspapers and posted to "thematic reflection" Padlets with ideas for our future digital history project, including format, purpose, and, as a repeat theme within digital history, access.

One theme that emerged from the analysis of student newspapers was campus traditions and community activities, many of which are still practiced on campus today. Many students read issues of *Revellations*, a student publication created "by the students and for the students of Revelle College," one of UCSD's seven colleges, between the mid-1970s and mid-1990s. The campus traditions described in the issues of Revellations from the 1980s and 1990s sparked thoughts about the power of nostalgia and community, especially in the context of COVID-19, when so many of us were away from campus. Final course reflections also touched on the significance of examining campus traditions through the newspapers. One student reflected, "Traditions are so important towards preserving a campus community. Maintaining those events are crucial to involving other students and giving them an opportunity they really wouldn't get anywhere else." They ended their reflection with a message for future students: "Get involved, participate, protest, do everything you can to be an active member of your Triton community."

As we thought about the best way to showcase the student newspapers

---

30. **LG**: And I personally never felt "stuck" because of the instruction, but rather creatively I wasn't sure where to go next. It was extremely helpful to see what questions were asked of me before and what questions were being asked in general to hone in on what we were doing as a class. And again—I wasn't doing this out of some sense of "oh we want to make sure this is cohesive!" but about following the lines of thought that I found interesting. Sometimes my next post would come out of what I didn't see other students talking about yet.

in a future digital history project, members of the class expressed their sense of belonging in the UCSD community, something they shared with both the authors of the student newspapers and future visitors to the project. Reading, analyzing, and thinking through ways to showcase the student newspapers in a future digital history project encouraged students to think of themselves as both current UCSD students and future alumni.[31] In their final course reflection, one student wrote, "Learning about...the student newspapers was awesome. I wish I'd known about them sooner because there are a lot of past insights held within them. I think it is also very cool that not only do you get to learn about history, but you get to do it from an alumni's perspective and it kind of hits home harder." Another student reflected, "I learned that our campus has a long history of students writing newspapers, and am glad to be a part of the reason why the tradition will continue."[32] In these quotes, there is both a reference to the past and a look towards the future, as students thought through ways to make these digitized items more accessible for other students and alumni to interact with them. In our final proposal, students suggested tools that would change the visitor experience of reading a static PDF to interacting with the newspapers, including timelines, polls, links to current campus resources, comments, and ways to submit suggestions for newspaper articles to add.

Similarities in the issues that were raised in historical student newspapers prompted students to both discover and reflect on the continuation of those issues today. For example, radical student groups of the 1970s voiced discontent over the disconnect between administration and students (UCSD Student Newspapers, 1972a, 1972b, 1973). In a comment on one of Logan's posts, one student noted that this "is still an issue that goes on at UCSD today with many of the organizations

---

31. **LG**: It felt like we would be taking the place of the students we were reading about. One day, future students will look at our digital history project, and build on it using the project itself as a primary source. Who knows what that will look like? Will it be alongside student newspapers written today? How will they understand digital literacy and access? It surprised me that this made me feel hopeful to think about myself as a part of UCSD's past—something for future students to interact with and maybe understand us. It's a connection with the campus that is not focused on networking and job-finding, but about understanding the moments in history we study.

32. **RJG**: I love that we have this space to document our thoughts! There is space for past, present, and future students (and also anyone who comes across the website) to interact with one another, and I think that's something unique that you can't always get outside of the internet. It feels like we are no longer only presenting a one-dimensional history where the past is the past, but it feels like now there is also an ongoing student narrative, where we are all in dialogue with one another in these digital spaces!

on campus." Rossel's analysis of the relationship between research on campus and the Department of Defense covered in one campus newspaper, *The Indicator*, taught other students about this long-standing connection (UCSD Student Newspapers, 1968). "I never fully understood how intertwined they were," one student commented. "I feel as if just as many students don't fully recognize the history of our campus." Another comment on Rossel's analysis read, "It's not something that gets a lot of mainstream attention despite how many students would probably be upset about it."[33]

In addition to the "letters to future students," we ended our proposal with a list of questions for past students who had worked on the newspapers, in the hopes of one day coordinating with alumni when the project is ready to launch. Many of those questions asked alumni to reflect on their experience of the UCSD campus community. For example, "Do you feel you made a truly significant contribution to either the campus or society in general during your time at UCSD?"; "Do you think UCSD has responded appropriately to the demands you've made in the past?"; and "Since we found your writings so interesting for the time period, did you ever feel that way when looking back at those before you?" Since the newspapers were vehicles for former students' expressions of political protest, identity formation, and campus traditions, reading and analyzing them allowed DigHist students to engage with alumni as both historical actors and members of the broader university community.

There was something significant about how analyzing the ways in which the experiences of past UCSD students opened up opportunities for current students to discuss their relationship to the campus community. One student wrote an analysis of a series of issues of *Revellations* from the late 1980s-mid 1990s where they engaged with one of UCSD's more unfortunate nicknames, "UC Socially Dead." In an article from the October 8, 1990 issue of the student publication, then-freshman student Jack Sharp wrote about the three "personalities" found among UCSD students: the "Tensies," whose "stress levels reach stratospheric heights every day," causing "any bio-feedback machine within miles," to "suddenly and inexplicably explode;" the "Partimaniacs," who "attempt to fit their classes in between parties and hangovers;" and finally the "Stressies," who could be "typically described as blurry streaks of

---

33. **LG**: I am not surprised that the typical class at UCSD doesn't utilize the student newspapers—there's a vested interest in not learning the history of UCSD if it isn't pretty or convenient for the administration to change. All we have left of George Winne Jr. (who self-immolated in protest of the Vietnam War) is a very minimal memorial. New and pretty buildings sell, more resources for marginalized groups do not—even if they're being begged for.

panic running through the Plaza (USCD Student Newspapers, 1990)." These classifications resonated with other DigHist students.[34] In their writing about Sharp's article, the student author noted, "After scrolling through the UCSD SubReddit every day for the past two years, it seems like almost everyone would be categorized into Tensies or Stressies, with so many posts highlighting a noticeable decline in mental health, social interaction, and connection to the campus community while attending college here." However, instead of focusing on stress and tension as the main takeaways from the past, they also urged other students to reconnect with campus traditions and other students. Instead of "UC Socially Dead," we should be "UC Socially Determined." One student commented, "I found your post not only inspiring and motivating, but it gets me so excited to be back on campus!" Another wrote, "Reading your post made me excited to go back to school and take advantage of the opportunities presented!" Reading about past students' experiences allowed current students to reflect on their own.[35,36] The digital format of both the DigHist course and the collection of student newspapers lent itself to this kind of reflection, as class members visited and revisited the historical sources and the feedback of their peers asynchronously. It was easy to see how some of the tools and methods

---

34. **MK**: As an instructor, the resonance of these categorizations in today's student population seems like such essential knowledge. I definitely did not assume that all UCSD students were "Partimaniacs," but I'm not sure how much I would have understood the (enduring) meaning behind "Tensies" and "Stressies" before introducing ungrading into my courses and inviting students to share what they thought about how grades have shaped their educational experiences. As a former UCSD graduate student, I was (and perhaps still am) a "Tensie," but it was helpful to hear how close to home these labels hit for others.

35. **LG**: I think I became really frustrated with the administration in a way I wasn't before, with the student newspapers teaching me that the on-campus problems I was already frustrated about were not new, and the administration still chose to do very little about it. I felt more community with those who were angry with the administration, but also, due to being unable to disconnect administration from the university, couldn't help but wonder what my schooling could have been like had problems from the 70s been solved in the 70s, rather than continuing on for the next generation to deal with. That isn't to say no progress has been made, but after reading the student newspapers, I'm hesitant to give any credit to the administration.

36. **RJG**: I think that depending on each students' positionality, the newspapers were received in many different ways. For me personally, as someone who was involved with campus organizations and even part of a department (Ethnic Studies) that was born out of student organizing on our campus and on other university campuses, these newspapers did not necessarily change the way I understood my own experiences as a student. Rather, they pushed me to consider how things have, or haven't changed, as I related the content of the newspapers to the contemporary moment. It also allowed me to take a step back and question the more technical side of these newspapers.

our class used to collaborate and maintain connection could be employed in a future digital history project.

Sharing Instructional Design

During the COVID-19 turn to remote learning, everyone, including students, took on the challenge of instructional design. The "structurelessness" of the DigHist course encouraged students (and the instructor) to be intentional about how—and why—each participant interacted with the material and with others in the class. This kind of agency can be uncomfortable and challenging and demands that all participants—including the instructor!—take risks.

In the mid-quarter reflections (seen only by the instructor and individual student), students provided positive feedback about ungrading and flexible deadlines, emphasizing how learners were able to "take charge of knowledge," in this class. However, there was also plenty of "digital discomfort" in the early reflections. As Kristin Allukian, Allison Duque, and Alexander Cendrowski wrote in their article on digital discomfort within the Suffrage Postcard Project, "The pedagogical framework of the project asks team members, through digital decentering, to challenge their willingness to invest in the research field, methodology, and co-researchers; to challenge power structures and dynamics inherent to almost any project; and perhaps most importantly, to challenge their sense of intellectual autonomy (or lack thereof)" (Allukian et al., 2020).[37] The digital discomfort that surfaced in the mid-quarter reflections related to the open-endedness of the final project, unfamiliarity with some of the tools that were new to some (Hypothesis, WordPress, and Padlet), and uncertainty with the flexibility of course deadlines. In other words, discomfort can arise when learners have more control over their own learning.[38] In response to the reflections, the instructor adjusted some of the formatting for distributing feedback, including creating places (a Google Jamboard and a Padlet) for feedback that applied to the collective work of the class, rather than for individual

---

37. The Suffrage Postcard Project can be accessed at https://thesuffragepostcardproject.omeka.net/about.

38. **RJG**: I think that this discomfort also has to do with all the unlearning that had to occur for some students. Based on my experiences, open-ended finals and flexible course deadlines were rare at UCSD, so perhaps a lot of students had to forget (or unlearn) about the structured finals and harsh deadlines that they were so used to. Additionally, the majority of the professors at UCSD opt to use Canvas as the primary digital space for their classes, so I think part of the discomfort also came from branching away from Canvas, which was overly familiar to many students.

students. Sharing collective feedback was done intentionally as a way to reinforce the value of sharing in-progress work, but also as a way of transparently communicating how feedback was developed.[39]

Decentering traditional classroom power dynamics by encouraging collaboration, sharing work in progress, ungrading, and embracing a final project that wasn't fully defined, helped to make the class more meaningful for all involved. The main idea behind the course was simply to generate ideas. There were no rules completely set in stone. The only things students were given for certain were the digitized archive of student newspapers and the use of the WordPress blog, Hypothesis annotations, and shared online discussion spaces. Understandably, that amount of freedom can be extremely disconcerting because the course was unlike many others students had previously taken. We were working towards a goal that we set, as a class, not one that had been set in advance by the instructor, the department, or the university. In one student's final reflection, they wrote, "I was hesitant to learn something new at first, but I am glad I did. This class was a lot of fun, especially learning while being an active participant." Sharing power was intentional instructional design, something that requires time for everyone to feel comfortable taking risks and establishing trust between participants. That trust turned into genuine interest and care about the direction of the project. Several students asked in their final reflections, despite the end of the class, "How can we keep track of this project?"

A course can not be fully "designed" before it begins. Every set of students has different needs and goals for their classes. Instructors (and instructional designers) need to take those needs and goals into account, sharing the responsibilities and power of course design between instructors, students, and staff (including archivists and librarians) before and during the course. If we want to hold students accountable for pursuing their education, then we also need to trust their input and form a collaborative learning structure beneficial to students first

---

39. **MK**: I had found a lot of benefits in sharing feedback for individual assignments in "public" (commenting on Padlet posts) because I could reference students' peer feedback in my comments and bring others into a larger conversation. But collective feedback also allowed me to share what I was thinking about the work that everyone was contributing to the course through their individual contributions. I also shared the links to the Padlet/Jamboard before I had finished all of my comments. It was feedback, in process, of in-process work. In previous courses, I had purposefully hidden all grades/comments until I had finished responding to everyone. Since the class was ungraded, there was much more flexibility (and much less anxiety) for me in terms of how I distributed feedback. I was also much more excited to read and respond to student work since I wasn't required to detract any points or assign a letter grade.

References

Allukian, K., Duque, A. & Cendrowski, A. (2020). Decentering digital discomfort. *Hybrid Pedagogy*. https://hybridpedagogy.org/digital-discomfort/

Association, A. H. (n.d.). *The career diversity five skills: Digital literacy*. https://www.historians.org/jobs-and-professional-development/career-resources/five-skills/digital-literacy

Putnam, L. (2016). The transnational and the text-searchable: Digitized sources and the shadows they cast. *The American Historical Review*, 121(2), 377–402. https://doi.org/10.1093/ahr/121.2.377

Rosenzweig, R. (2006). Can history be open source? Wikipedia and the future of the past. *Journal of American History*, 93(1), 117–146. https://doi.org/10.2307/4486062

UCSD Student Newspapers. (1968). *The Indicator*. Special Collections & Archives, UC San Diego. https://library.ucsd.edu/dc/object/bb11956213

UCSD Student Newspapers. (1972a). *Black Voices*. Special Collections & Archives, UC San Diego. https://library.ucsd.edu/dc/object/bb55983863

UCSD Student Newspapers. (1972b). *Lumumba Zapata*. Special Collections & Archives, UC San Diego. https://library.ucsd.edu/dc/object/bb2905731p

UCSD Student Newspapers. (1973). *Nine*. Special Collections & Archives, UC San Diego. https://library.ucsd.edu/dc/object/bb8635959n

UCSD Student Newspapers. (1990). *Revellations*. Special Collections & Archives, UC San Diego. https://library.ucsd.edu/dc/object/bb90113897

Feeling (Un)Seen
Notes on hidden disabilities in the (digital) classroom

Andrew David King

> Loving intensives of Intensive Care
> Bear down on your given name,
> Margaret, attend, attend now
> Margaret, they call you to live intensely
> At the moment of your medication.
>
> –Josephine Miles, "Intensives"

> What did one do with the body in the classroom?
>
> –bell hooks, "Eros, Eroticism and the Pedagogical Process"

1

The pivot to virtual teaching in the COVID-19 pandemic—the details of which are still being ironed out, nearly three years later, at educational institutions worldwide—seems like an especially good time to reevaluate how the concept of social legibility and the related concepts of privacy, care, and obligation might inform our pedagogical theory and practice. As instructors, how do we interpret and understand our students within the bounds of our professional and ethical obligations to them? How do our students, in turn, "read" and understand us whether behind desks or screens? What institutional prerogatives, codified or uncodified, must our students contend with and respond to simply in virtue of showing up to class? What habits of discourse, dress, speech, and thought do we, often without realizing it, impose on our students—and which of these habits serve no deep educational purpose or, worse, actively hinder their development?

To these important, often overlooked questions, I want to add a few more. What can a careful consideration of hidden disabilities—I prefer the plural term, as a way to resist the linguistic flattening of diversity within the concept of "disability" itself—reveal about the legibility of students' lives in the classroom? More importantly, what can our students with hidden disabilities, individually and collectively, teach us about how to teach them?

The term "hidden disabilities" is, at best, imperfect. These disabilities aren't hidden to those who have them, and often they aren't meaningfully hidden at all; others, for various reasons, fail to notice them. Despite thoughtful criticisms of the term, however, I believe it has useful applications and I claim it for my own experiences as a disabled person. Clearly not every aspect of one's identity is capable of being inferred by others on the basis of one's appearance or behavior. Many disabilities that aren't truly hidden become hidden by others' shortcomings in knowledge, education, or sensitivity. There are other disabilities—forms of chronic pain or episodic impairment, for instance—that simply aren't revealed unless the disabled person in question volunteers the information. As I'm using the term, then, it has at least two applications: to refer to instances where a disability is obscured for social reasons, and to refer to instances where others, no matter how sensitive, couldn't know about it unless told.

There are, then, questions of how to think about hidden disabilities in the classroom in general, and questions about how these considerations, taken seriously, naturally lead us to modify our teaching. To these I'll add a final concern: what are the challenges—and the unsung advantages—of virtual teaching for students with hidden disabilities? Even as schools and universities return to "normal," it seems a safe bet that virtual instruction will persist with virtual attendance options offered for disabled and ill students with increasing frequency. We—especially the able-bodied and healthy among us—should think about what virtual instruction offers students whose experiences are unlike our own, and strive to retain what's valuable from it as institutional inertia drags pedagogy back to a pre-pandemic state.

An abstract discussion like this one can't but fail to do justice to the particularities of students' experiences, but I offer these reflections as a preliminary sketch, a map for reconfiguring our attitudes, expectations, and strategies. These notes reflect on the nature of disability, "hidden" and not, and my time as a student and instructor, both in-person and virtually, during the pandemic. My recommendations straddle the divide between the theoretical and practical. My thinking and experience have led me to endorse not a sleek package of best practices, but what might be described as an ethical disposition: towards the idea of teaching, towards our students, and towards the communities, both concrete and ephemeral, of our classrooms.

2.

We unthinkingly think of disability as a legible thing—something available for inspection. Lennard J. Davis, writing about this cultural assumption notes that "[t]he person with disabilities is...brought into a field of vision, and seen as a disabled person" (1995). Although one can begin to understand what I mean by "legible" here by exchanging the term for "visible," to equate the legible with the visible is to simply push the problem back. For something's being legible, on my understanding, isn't a matter of its being available to some perceptual faculty, like sight or hearing, but a matter of its being available to be interpreted as something meaningful within a given social or cultural context.

The concept of legibility, dealing as it does with the means by which persons and groups understand each other, figures in the disciplines of history, social and cultural theory, and psychology among others. Michel Foucault, writing of working-class housing estates in the nineteenth century, describes how the estates' layouts made individuals "visible," and how "the normalization of behavior meant that a sort of spontaneous policing or control was carried out by the spatial layout of the town itself" (1997, p. 251). James C. Scott, in *Seeing Like a State,* details how early modern European statecraft seemed intent on, "rationalizing and standardizing what was a social hieroglyph into a legible and administratively more convenient format" (1998, p. 3). The American Psychological Association's online dictionary specifies, as part of its second definition for "legibility," that the term refers to "the ease with which an environment can be cognitively represented," among other things. Relatively recent empirical work on the concept has explored how members of different cultures variably experience a given environment as socially legible in this sense.[1]

Most of the sources mentioned so far discuss legibility as a higher-level property of social arrangements and systems, but I'm mainly interested in the ways in which persons—students and instructors—can be legible to each other. Consider a more down-to-earth example that relies on visual perception. Suppose someone wants to go to a Pride parade to show support for their friend who's marching in it. In order to make clear that they're attending in support of the parade, this person might wear rainbow colors or pins that indicate their celebration

---

1. See the American Psychological Association's online dictionary entry for "legibility," https://dictionary.apa.org/legibility, and, e.g., "Social Legibility, The Cognitive Map and Urban Behaviour," Thierry Ramadier and Gabriel Moser, *Journal of Environmental Psychology* 18, Issue 3, September 1998: 307-319.

of Pride. In doing so, they're rendering themselves legible as a Pride supporter to their peers—and to everyone who understands what these symbols mean in broader cultural contexts. Or consider the phenomenon of being called by one's given name, as in my epigraph from Josephine Miles. To be called by one's given name is to be read as a particular person, one with a particular historical existence.

Legibility can also have a nefarious side. Frantz Fanon's writings about race, racism, and colonialism in *Black Skin, White Masks* include numerous traumatic recollections of being called slurs—being socially read and addressed as subhuman or inhuman. In addition, Fanon painstakingly outlines the phenomenology of being read as representing not just oneself but one's perceived culture or race. Recalling how he was often left two or three places on a train, Fanon describes how he "existed triply"—how European colonial society made him accountable for his body, his race, and his ancestors (1986, p. 112). Fanon's observations issue an important qualification to what others, like the philosopher Charles Taylor, have argued is the indispensably dialogical character of identity formation (1994, p. 33). Fanon notes that this dialogical relationship can just as well be oppressive, as when the social atmosphere is so charged with racism that what he calls the dialectic between body and world is definitively structured by it (1986, p. 111). Another example of ethically troubling legibility is discussed by Jesus Cisneros and Julia Gutierrez in their writing about what they term "undocuqueer" experience (2018). For them, legibility is linked to state legitimation as expounded by Judith Butler, who claims that the latter involves finding that "one's public and recognizable sense of personhood" fundamentally depends on prescribed discourse (2004, p. 105). Lastly, while failures of legibility can lead to bad consequences, it's also possible, as Alison Kafer notes, for the needs of the disabled to be problematically made "hypervisible," framed misleadingly as exceptional and contentious (2017, p. 215).

A decade into several chronic illnesses, I still find myself in the throes of thinking of disabilities as necessarily legible. Consider the International Symbol of Access: the stick-figure nested in a swooping curve, a symbolization of a wheelchair user, that's printed on signs and spray-painted on parking lots internationally. Rejected attempts at revising this symbol have proposed a stick-figure body that appears to be in motion, rising from the chair or moving with it. But as critics have pointed out—echoing critiques of the original symbol when it debuted—this modification to the icon, whatever its merits, still failed to address the question of how to include in a mainstream conception of disability those whose disabilities aren't readily perceived by others.[2] Among the latter, we might include individuals with mental illnesses or learning disabilities, or those with chronic pain, fatigue, or illness, besides many more. How might the able-bodied "see" those with hidden disabilities in such a symbol?

The short answer, of course, is that they can't. Common cultural understandings of disability start with what is, for better or worse, a familiar case of disability—that of the visibly-identifiable wheelchair user—and imaginatively extrapolate from there if they extrapolate at all. My claims here are intended descriptively, not critically, though there's plenty to criticize about these unreflective understandings of disability. At the same time, there's something to be said in favor of the strategic essentialism that the International Symbol of Access promotes. I think here of my mother's witnessing someone being written a hefty ticket for parking in a disabled spot. Impatient with the woman's protests that she'd only meant to be there "for two minutes," the officer said, "Ma'am, this is for people who can't walk." If that retort was false to the letter, it was true to the spirit.

Where legibility gets tricky in the classroom is when it involves students with hidden disabilities forfeiting their privacy in order to prove, formally or informally, that they have a disability and therefore deserve a certain accommodation. Whether these individuals are ever "temporarily able-bodied" or not, their disabilities don't readily present themselves to public inspection or verification. While there's a simple sense in which we can see (or hear, or feel) that someone uses a wheelchair to get around, there's no correspondingly simple sense by which we can perceive that someone has a hidden disability unless they tell us. There might be exceptions to this general characterization: someone with chronic, recurrent pain might be forced to sit down and take medication, which could then be publicly interpreted as a sign that they live with an invisible disability.

But these exceptions are also subject to qualification. Someone's suddenly expressing pain and seeking to resolve it could just as well be interpreted as something temporary, a cramp or headache, rather than a persistent condition. The symptoms of those with hidden disabilities can also be time-dependent; rather than acting as reliable public signposts of a disability, they may come and go. Even worse, as I've noted, those with hidden disabilities often find themselves, as N. Ann Davis attests, in the position of having to prove the existence of their disabil-

---

2. See, e.g., Emma Teitel, "Critics of new 'dynamic' disability symbol not just anti-PC cranks," Toronto Star, 2017

ities to others; this, and the mistrust with which such demonstrations are sometimes received, incurs emotional and psychic costs (2005). Linda Martín Alcoff, writing about race and gender as visible phenomena, observes that "if there is no visible manifestation of one's declared racial or gendered identity, one encounters an insistent skepticism and an anxiety" (2006, p. 7). Something similar is true, I suggest, for disabilities and illnesses.

My use of the term "read" in relation to social legibility isn't meant to imply that individuals with hidden disabilities are texts, or that they exist for, or because of, our interpretation. It's meant to draw an analogy between reading texts and what we do in social situations—like teaching or being in a classroom—where we make inferences about others in advance of, and in addition to, their own testimony about themselves. In my view it's a pervasive, inescapable, and not necessarily nefarious phenomenon. Just as we can't help discerning that gathering clouds and a darkening sky might mean rain, so we can't help discerning that, say, the person standing before us may be a fan of a local sports team if they're wearing a shirt that suggests this. Note that volition, while compatible with legibility, isn't a part of the concept. When we talk about what's socially legible to us, we're talking, at least in part, about what information about other people "shows up" to us without our consciously intending or doing anything at all. Who and what is legible to us, and how, may have to do with what Ludwig Wittgenstein called our "form of life," our "Lebensform," a set of background presuppositions intimately related to our store of linguistic terms and concepts (2009, p. 19).

3.

At the several universities where I've taught, students are thrust into context after context where their academic success—a different metric, I suggest, from their learning—is dependent on their rendering their thoughts, speech, and bodies legible to their instructors and to the institution. In addition, these institutions address them, at different times and according to their convenience, both as subjects and as customers.

There's no getting around a certain amount of legibility; we have to know, and for good reason, what the names of our students are and we should want to know what they want to be called and what pronouns they use, if they wish to offer them. What I've been calling "social legibility" has a scholarly analogue, too. From the perspective of progress within a discipline, new work, however innovative, has to be interpretable in light of what preceded it. And certain classroom exercises—think first-day-of-class icebreakers—that unfold in the spirit of friendship, community, and mutual understanding can be considered legibility-building. Where an emphasis on legibility becomes troublesome is, again, where it forces students to redesignate private parts of their lives as public.

When we structure our courses with what Rosemarie Garland Thomson calls the "normate," maximally non-disabled student in mind, we risk forcing our disabled students, including those with hidden disabilities, to choose between passing up accommodations or making legible, on the university's terms, aspects of their life they may not wish to share (1997, p. 8). In order to better understand why this should be concerning, consider first what many students go through to secure general disability accommodations.

At each university at which I've either studied or been employed, the process of obtaining disability accommodations requires that a student demonstrate that they have a disability or medical condition for which a certain set of institutional remedies has been authorized. An obvious but under-discussed observation is that such a policy will have classist ramifications in any country that lacks public healthcare: intended as a safeguard against abuse, asking for proof disenfranchises those unable to afford it. Well-off students with good doctors and medical plans will likely have a diagnosis in hand before stepping on campus, while the less well-off will be forced to spend time, money, and labor navigating purposefully Byzantine systems to obtain basic statements (or restatements) of the facts of their situations. They might have to do this repeatedly throughout their education, as some universities set their own expiration dates on students' medical documents, even if those documents pertain to lifelong conditions—a practice that, as far as I can tell, has no medical justification. The documents, once submitted, go through opaque review processes where it can be unclear who has access to them and what the terms of that access are. The student may have to participate in interviews and appointments with university staff before the accommodations are granted. All this has highly worrisome implications. As a recent study in *Disability Studies Quarterly* suggests, a student's success in obtaining accommodations, or "learning supports," may depend on how well students manage to navigate their university's bureaucracy and how lucky they are in finding supportive, responsive faculty (Bruce & Alward, 2021).

Note, here, one assumption of disability accommodations processes: that students can be placed into mutually-exclusive "disabled" and "non-disabled" categories from the point of view of the university. (I'll continue to speak of "the university" in the abstract, as a stand-in for what numerous institutions of higher education share.) Or, if that's too metaphysically loaded, "students eligible for accommodations" and "students not eligible for accommodations." Practical binaries like these serve a corporate-bureaucratic ideal of efficiency. Students are sorted, using the rhetoric of compassion, into two sets, one with respect to which the university becomes obligated to expend more of its resources.

What such a binary doesn't do is help students whose disabilities or illnesses are difficult to diagnose, uncommon or poorly-understood, or not even unambiguously acknowledged as disabilities or illnesses. Consider those who live with chronic pain, chronic fatigue syndrome, fibromyalgia, or simply an undiagnosed ailment whose symptoms have no locatable cause or which have outlasted their expected duration. If a physician's note attesting to the fact that they believe the student to be living with a medical disability is sufficient to secure accommodations for the student in the absence of a formal diagnosis, why not just believe the student? Students in these situations might decide, not unreasonably, that the labor of applying for accommodations—of finding a doctor sympathetic enough to testify on their behalf—is simply not worth it. I myself made this decision more than once.

Students with disabilities are faced, then, with a dilemma. Either render themselves and their challenges legible to the university on its terms, or forego accommodations. For students with conditions, like chronic pain or fatigue, already barely legible to the medical establishment, the stakes are even more dire, as they may do everything right and still fail to be seen. Their knowledge of their own body, of their own experience, may be disqualified—classed as one of what Foucault terms "subjugated knowledges," or modes or articles of knowledge that are either "masked in functional coherences or formal systematizations" (notice how this masking is closely related to, if not identical with, illegibility) or discounted as knowledge (1997, p. 7). They become victims of what Miranda Fricker calls "hermeneutical injustice," where the limitations of a culture's social resources preclude a group or individual from articulating their experiences or even fully understanding them (2007, p. 1).

When we take stock of the fact that disabled students face choices like this before even stepping into our physical or virtual classrooms, we can begin to appreciate the ethical stakes of accessible course design. We start to see that our obligation to conscientiously design our courses doesn't end with the inclusion of a well-intentioned rider on our syllabus about disability accommodations—often part of mandatory departmental boilerplate passed over in a class's first meeting—but begins there.

4.

I speak as a student and instructor whose life is marked by several such disabilities, some mental, several physical. Though I feel literal pain—the chronic pain that has come to characterize my day-to-day existence—as I write this, something in me still worries I'll be called out as an impostor. I don't use a wheelchair. I'm occasionally, though increasingly less so, what some call "temporarily able-bodied," or asymptomatic enough to pass as, and trick myself into thinking I am non-disabled. I don't have a disabled parking placard, or many of the other visible trappings of disability. But I'm every bit as dependent on the support of other people and institutions for my well-being as the prototype depicted in the International Symbol of Access.

In fact, in 2021, I may be more so as my conditions require constant check-ups, monitoring, and interventions, none of which I'm guaranteed in a country without public healthcare. Whenever I go out, I think about how I'll be able to access emergency medical services, and take a pouch of medicines with me. Every two weeks, I painstakingly organize a rainbow of pills into 28 clear plastic squares, each representing either a morning or an evening dose for the next two weeks. I keep a journal of symptoms in an attempt to puzzle out the tea leaves of my pain.

I recite this abbreviated personal history not to garner pity, but as a spell against denial. Even now, my social conditioning causes in me a kind of double vision about my own body. In the bathroom mirror, I see a youngish, male-presenting torso that visually registers as "healthy." How could this body be sick? How could the owner of this body—indulge my dualism here for a moment—feel scared, terrified, even, by what's happening to, or inside, it? How can the medical "reading" of my body as "mostly healthy despite several ongoing conditions" be reconciled with my own day-to-day experience of pain, discomfort, fatigue, frustration, alienation, and isolation? And if I can hardly formulate these questions for myself, how can I hope to meaningfully communicate about them with others, whether in order to secure the accommodations I need or to find my fellow travelers? Should I carry a cane with me, as I've often considered doing, not because I need it, but to signal to the able-bodied that I'm not one of them?

Lacking these answers, I've been led to drastically reevaluate what it is that I can reasonably ask of my students. I mean "ask" not only in the sense of action, or what one might ask someone else to *do*, but of information. What can I reasonably expect my students to know—about themselves, about their conditions and disabilities—and what can I reasonably ask them to share with me?

5.

No individual educator will be able to rewrite their culture's conception of disability or solve all the problems entailed by their institution's Procrustean disability accommodations process. But individual educators can choose to enact more expansive conceptions of care, of legibility, that aid students with hidden disabilities while respecting professional boundaries.

Some disabled students will very much need services that only the university is able to provide. I think here of funded note-takers, for instance. I'm not recommending that students eligible for accommodations not apply for them, nor that instructors intervene in conversations between these students, their doctors, and administrators. Nor do I want to disregard the ways in which university disability services can sometimes help disabled students to become legible in ways that are useful for them: by serving as a form of proxy legibility while adjudicating correspondence about accommodations between students and instructors, for instance.

What I'm recommending is that, as much as possible, we not put our students in the position of having to make themselves, and their disabilities and illnesses, legible to us or the university. Note that I say "having to make": not forcing students to become legible isn't the same thing as forcing them to remain illegible. I try to find subtle ways to let students know that I'm available to discuss, within professional boundaries, how their challenges might be affecting their performance in class. I try to signal to my students that I live with illness and disability, and that openness about one's physiological, medical, or mental state is neither a faux pas nor a "trauma dump."[3] When a student

---

3. This is not a risk-free endeavor. Besides a general disdain among academics for perceived weakness, of which talking about one's disability or illness can be taken as a sign, students can also hold these outdated attitudes. I once

mentions dealing with illness, or needing to take a day off for mental health, I imply I've dealt with similar things. I mention in class how essential it is to practice self-care at every level of academia, making clear that I make time for this care in my own life, too.

Universities already protect students from being asked about the specifics of the disability or condition for which they're receiving accommodations.[4] But what if we, as instructors, took a different approach to course design, one that doesn't tacitly assume that all our students are able-bodied, and that if they aren't, they'll seek formal accommodations? What if disabled students, and not non-disabled students, were the targets of our curriculum?

These kinds of imaginative suggestions tend, in my experience, to be brushed off as naïve. To seriously entertain relaxing or removing in-person attendance requirements, to make flexibility about deadlines (even within reason and the terms of the instructor's contract) the rule and not the exception, to offer alternatives to verbal contributions to class discussion: to a certain cast of mind, these approaches are pedagogically suspect because they are not "rigorous" enough. In response, we should consider how appeals to rigor can have more to do with the enforcement of social norms than with any genuinely intellectual goals. David Bartholomae, in "Inventing the University," has written sensitively about the complex demands that academic writing pedagogy places on students to make themselves legible, where the notion of legibility at work can sometimes be less a matter of making intelligible to a particular audience who one already is and more a matter of repackaging, even reshaping, oneself and one's experiences for this audience (1986). Even taking appeals to rigor at face value, it's hard to divorce them from educational models that tie learning to capitalist notions of productivity. As David Wasserman argues, "Much of the accommodation students receive as a result of a diagnosis of disability should not require one; perhaps more flexible exam times

---

received a course evaluation from a student who found it "borderline inappropriate" that I mentioned, in an aside one class, that I was dealing with the side-effects of a new medication that day, and so asked for their patience. But whatever my fears of the consequences of making my own disabilities and illnesses legible to my students, those fears aren't my priority as an instructor. (There's much more to say here, of course, about the safety and security of disabled faculty, especially those without job or healthcare security, and my setting them aside is not intended to downplay them.)

4. An FAQ regarding disability accommodations at UC Berkeley, my alma mater, states: "Students with disabilities have a right to privacy regarding their medical diagnoses and medical documentation. For this reason, we ask faculty not to request medical documentation from students with disabilities."

and course requirements should be more broadly available; perhaps our school systems should [be] oriented less toward performing a screening function for employers and more toward providing an 'appropriate education' for *all* children, not just disabled ones" (1998).

I don't pretend to have the experience and insight necessary to give both meaningful and general prescriptions about how to construct courses so as to allow students maximum freedom with respect to their own legibility. I can only urge a conception of the role of instructor that, as much as possible, respects students as unknown quantities—one that doesn't assume, in particular, that they're able-bodied or that they don't have hidden disabilities. As an instructor, I've worked directly with students to help them be successful when institutional guidelines had little or nothing to offer them. Though these arrangements may not have been legible to university administrators, there was nothing suspect about them. They involved no crossing of any boundary, no violation of any policy that went beyond my own prerogatives. They didn't involve anyone revealing information they didn't want to reveal, or putting another person in a position of liability. Several of these students confided to me that they had been either been turned down from university disability services, or hadn't received the accommodations they needed most. I was their last link to institutional accommodations.

Accommodations need not always be spoken of as stopgap measures. Sometimes, the flexibility I offered students—more time to complete assignments during a flare-up of symptoms, for instance—positively aided their intellectual progress, more than simply not interfering with it. By permitting them to spend fewer of their able-bodied hours performing a hyper-legible "student" role, they could spend more time actually being students in ways that worked for them.

6.

If one set of problems related to legibility presents itself in physical classrooms, another arises in virtual ones. The body language and facial cues that we typically rely on can become distorted or effaced. The gridded arrangement of faces on Zoom can thwart the natural information-overload-prevention device of focusing just on whoever's speaking or whoever's closest to us.[5] And yet, I want to suggest these modifications aren't fatal. In fact, we can draw lessons from them

---

5. See Jeremy N. Bailenson, "Nonverbal Overload: A Theoretical Argument for the Causes of Zoom Fatigue," in *Technology, Mind, and Behavior*, 2021.

about legibility and student accommodations that we can then apply to in-person teaching.

In March 2020, when the university at which I was teaching decided to transition all in-person courses online, I made my course—an advanced seminar on modern poetry with a workshop component—entirely asynchronous. We wouldn't meet online at our scheduled time; we wouldn't pretend like the pandemic wasn't wreaking havoc on our lives, as if the only change it involved was our not being able to meet in the same room. Thanks to supportive supervisors, I could do this, but my colleagues in another departments couldn't. There, an ableist conception of "rigor" ruled the day, one that not only demanded synchronous teaching but an increase in difficulty to compensate for the perceived loss of intellectual integrity that virtual teaching entailed. Never mind that, at the same time, students were dealing with the fallout from sick or dead relatives, from their parents' losing their jobs, or from having to suddenly care for younger siblings who couldn't attend school. A colleague in that department reported that a few of her students simply went radio silent, as if they'd vanished.

With asynchronous assignments, my students were able to utilize whatever pockets of calm and clear-headedness they could find—in short supply, no doubt—to complete the course. The major hurdle was figuring out how to conduct a writing workshop where students' poems were discussed and critiqued, while all of us were in different locations in time as well as space. I resist saying "replicate" a writing workshop; what we manifested together wasn't a poor facsimile of a genuine article, but a real form of community under emergency conditions.

Rather than asking what my students lost in my class's moving online, I want to ask what they might've gained. Counterintuitively, I found that the students on the whole did well, shock and stress of the pandemic aside. Some students' performances improved; several made substantial contributions to workshop discussions in writing that I suspect they would've never had made verbally. For my more rural students, and for myself, foregoing commuting meant gaining time and well-being. I've since hoped that the pandemic would spur more discussion about whether requiring attendance risks discriminating against disabled and ill students, or at least making their lives markedly harder, but there's been a sorrowful paucity of conservation about this. Students with hidden disabilities may be "temporarily able-bodied," or free of mobility issues. But they may also be students for whom energy is in short supply. Students for whom a daily commute is highly taxing and represents lost time that could have been better spent studying or recuperating. Of these students, some, due to their disabilities and conditions being handled better at home, won't gain anything from physically attending classes; some who might in fact gain from them will have those gains mitigated or canceled by the travails of commuting. Here I point again to my own experience as evidence. But recently more general empirical evidence has been gathered that supports the claim that commuting time is inversely related to good health.[6]

In synchronous online classes I've taught since, Zoom's various functions beyond the audiovisual have provided a virtual body language, a substitute by which students who might not have been able to use, or might not have preferred to use, physical body language and speech availed themselves of its nuances. Students used the "raise hand" feature to note their desire to speak, sparing them the anxiety of having to intervene in fast-paced discussions. The ability to virtually and visually "raise one's hand" at any time, in a way made non-awkward by its standardization in the Zoom platform, helped those silenced by anxiety and by social practices inhospitable to their communication styles, to be perceived as speakers. Students who needed to turn off their webcams—whether because of a disability or illness, to take a break from social participation, or for some other reason—did so, reappearing visually only when this was manageable or desirable for them. Most of these students did well in the course, and several were frequent participators.

Zoom's virtual chat has also figured prominently in my experiences with online teaching. The chat not only served as a forum where those who wished to contribute to discussion in writing could do so, but as a site of free encounter, discovery, and conversation. Although a flurry of chat activity can be distracting, the chats in my virtual classrooms almost always ended up serving as crucial venues for questions, answers, reactions, rejoinders, and objections. It's tempting to view the chat skeptically, and I can recall a few instances where it hosted problematic behavior. But the risks involved are similar to those involved in verbal discussions, where students can also say disruptive, offensive, or off-topic things. What matters, I suspect, is the framing. At the start of each class I teach, I set community norms with my students. The calls for basic respect and courtesy that these norms inevitably include are understood, in virtual classes, to encompass the chat. I make clear that I'll be dipping into the chat occasionally, but not policing it, treating it as a space for "hallway conversation": nothing said there should vio-

---

6. See, e.g., "Tired of Commuting? Relationships among Journeys to School, Sleep, and Exercise among American Teenagers," by Voulgaris et al., in the *Journal of Planning Education and Research*, 2017.

late our norms, but I won't be inspecting it, though I'll try to respond to questions directed at me.

Those tempted to think of the chat as a site where mere anarchy is loosed should reconsider the enforced silence of the traditional lecture hall, where students who might wish to discuss the lecture as it unfolds are prevented from doing so. Instead, they're faced with the task of absorbing new information while having to physically signal their engagement to their instructor and peers via body language, even if this latter task detracts from the former. As pedagogical theorist Martha Burtis pointed out to me, there's no guarantee that the performance of engagement tracks actual engagement. Perhaps as instructors, we've become too fixated on feeling like we're being paid the attention we deserve. But I would much rather my students dynamically, even disruptively, respond to what I have to say than sit in siloed silence, together but also alone with their peers.

In my experience, student use of the Zoom chat alongside lecture or discussion didn't preclude more traditional academic forms of engagement. What's more, no one who found the chat uninteresting or not useful had to read it, though I grant that the situation would be improved if individual users could turn off chat notifications. Micro-communities developed in those chats, with regular contributors casting themselves as characters: the objector, the helpful answerer, and the wit, for instance. Somehow, surprisingly, this lively, thriving academic culture, before relegated to the hallways outside of class, had been made legible in its own way.

7.

What does this all mean for students with hidden disabilities? I recount my experiences with online teaching in the pandemic to shore up my sense that the prospects of virtual teaching are more optimistic than one might think, especially for disabled students. Virtual classrooms hold out the promise of more flexible, more accommodating models for a greater variety of students and learners; in the care of the right instructor, they allow students increased freedom in how they render themselves legible or illegible. Freedom in this sense is a boon, most of all, to those who lives would be made most difficult without it—a group that includes students with disabilities, hidden and not.

These benefits come in tandem with all the added challenges of optimizing new technologies, which I don't want to downplay. Still, those who taught online during the pandemic should embrace the opportunity to return to in-person teaching, if and when they do, with more inclusive pedagogical practices gleaned from the much-maligned virtual classroom.

To put the point more forcefully: I'm convinced that educators who made it "through" the pandemic (which is, in fact, still ongoing) only to adopt as soon as possible all the same pedagogical methods they had before it have thrown away a crucial learning opportunity—the opportunity, literally, of a lifetime. For American instructors, whose students don't enjoy the social safety nets of healthcare or living wages, the risks of failing to learn these lessons are especially serious. And yet the rhetoric of a return to "normal," and of the assumed superiority of non-virtual teaching, persists.

8.

During the U.S.'s grappling with the pandemic, anti-public health apologists have often employed the rhetoric of "protecting the vulnerable" —the immunocompromised, the elderly, the disabled—in the service of resisting lockdowns.[7] These calls were often contentless, shifting the responsibility for dealing with some of the pandemic's worst effects onto those most likely to suffer from them. But the sentiment's core idea—that, in thinking about social arrangements, we should think first and foremost about those made worst-off under them—has a storied history, one that deserves to be retrieved from these more recent, bad-faith appropriations. The political philosopher John Rawls's foundational work *A Theory of Justice* makes considerations pertaining to the lives of the worst-off members of a society the cornerstone of its titular theory.[8] I want to conclude these speculative notes by thinking about the potential of such a focus to transform pedagogy, online and offline.

I don't wish to claim that the students whose sufferings and challenges are interpersonally and institutionally illegible are necessarily those who ought to be considered the worst-off when we think about our

---

7. Coverage about the so-called Great Barrington Declaration, which involved scientists from the U.S. and worldwide, focused on such rhetoric, which Declaration supporters often employed. See, e.g., "Coronavirus: Health experts join global anti-lockdown movement," *BBC*, 2020, https://www.bbc.com/news/health-54442386.

8. Characteristically, Rawls holds, e.g., that "social and economic inequalities, for example inequalities of wealth and authority, are just only if they result in compensating benefits for everyone, and in particular for the least advantaged members of society" (*A Theory of Justice*, 13).

classrooms. Some students face major obstacles from conditions and disabilities that are very legible, in the sense I've been describing, both to individuals and institutions. But insofar as our pedagogy fails to address, and so disadvantages, the particular needs of any student, that's an extent to which we should hope to revise it.

The necessary revisions won't always or even often be possible; they may be thwarted by the limits of our own labor, by bureaucracy, or by other limitations foisted upon us.[9] What isn't subject to material or administrative limitations, however, is our ethical imagination, which includes the most basic means by which we interpret and evaluate others. Adopting a maximally able-bodied approach to course design forces students with disabilities to undertake the often-painful work of making themselves legible to normate culture, which already excludes and erases them. The labor of making themselves legible in the classroom and in the university becomes, then, a second insult. No doubt we can't, as I've noted, prevent all disabled students from undergoing every kind of institutional maltreatment, or from ever having to render themselves legible. Nor should we necessarily want this, as legibility itself is a neutral phenomenon—part of what it means to be a social being, as I understand that notion. But we can expand the boundaries of our concept of the "student," focusing more on what we can ask of our students as intellects, thinkers, and writers than as bodies with differing, contingent needs. And we can focus on getting clear, with ourselves and our colleagues, what it is that we believe about instruction without quite realizing that we believe it, and what in these unexamined beliefs reinforces ableism in academia. As the rhetorical theorist V. Jo Hsu writes, "Only by rendering our beliefs legible—by fixing them temporarily, imperfectly, and vulnerably in some form—can we create the conditions for discussion, negotiation, and change" (2018, p. 163). This requires bravery and risk, in quantities our disabled students already know much about.

References

Alcoff, L. M. (2006). *Visible identities: Race, gender, and the self*. Oxford University Press.

Bailenson, J. N. (2021). Nonverbal overload: A theoretical argument for the causes of Zoom fatigue. *Technology, Mind, and Behavior*, 2(1). https://doi.org/10.1037/tmb0000030

---

9. See, e.g., Martha Burtis's critique of Quality Matters in "The Cult of Quality Matters," *Hybrid Pedagogy*, August 10, 2021, https://hybridpedagogy.org/the-cult-of-quality-matters/.

Bartholomae, D. (1986). Inventing the University. *Journal of Basic Writing*, 5(1), 4-23.

BBC. (2020). Coronavirus: Health experts join global anti-lockdown movement. https://www.bbc.com/news/health-54442386

Bruce, C., & Aylward, M. L. (2021). Accommodating Disability at University. *Disability Studies Quarterly*, 41(2).

Burtis, M. & Stommel, J. (2021). The Cult of Quality Matters. *Hybrid Pedagogy*. https://hybridpedagogy.org/the-cult-of-quality-matters/

Butler, J. (2004). *Undoing gender*. New York. Routledge.

California, U. of & Berkeley. (n.d.). Berkeley Disabled Students' Program Faculty Frequently Asked Questions. https://dsp.berkeley.edu/faculty/faculty-faqs

Cisneros, J. & Gutierrez, J. (2018). "What Does It Mean to Be Undocuqueer?" Exploring (il)Legibility within the Intersection of Gender, Sexuality, and Immigration Status. *QED: A Journal in GLBTQ Worldmaking*, 5(1), 84-102.

Davis, L. J. (1995). *Enforcing normalcy: Disability, deafness, and the body*. Verso.

Davis, N. A. (2005). Invisible Disability. *Ethics: Symposium on disability*. 116(1), 153-213

Kafer, A. (2017). Bodies of Nature: The Environmental Politics of Disability. In J. Ray & J. Sibara (Eds.), Disability studies and the environmental humanities: Toward an eco-crip theory (p. 201-241)

Foucault, M. (1997). *"Society must be defended": Lectures at the collège de France, 1975-1976*. (D. Macey, Trans.) Picador. (Original work published 1976)

Fanon, F. (1986). *Black skin, white masks*. (C. L. Markmann, Trans.) Pluto Press. (Original work published 1952)

Fricker, M. (2007). *Epistemic injustice: Power and the ethics of knowing*. Oxford University Press.

hooks, b. (1993). Eros, eroticism and the pedagogical process. Cultural Studies, 7(1), 58-63. https://doi.org/10.1080/09502389300490051

Hsu, V. J. (2018). Reflection as Relationality: Rhetorical Alliances and Teaching Alternative Rhetorics. *College Composition and Communication*, 70(2).

Miles, J. (2011). Intensives. In S. Black, J. Bartlett & M. Northen (Eds.), Beauty is a verb: The new poetry of disability. Cinco Puntos Press.

Rawls, J. (1999). *A Theory of justice, rev. ed*. Belknap Press.

Scott, J. C. (1998). *Seeing like a state: How certain schemes to improve the human condition have failed.* New Haven, Yale University Press.

Taylor, C. (1994). *Multiculturalism: Examining the politics of recognition.* Princeton University Press.

Teitel, E. (2017). Critics of new "dynamic" disability symbol not just anti-PC cranks. Toronto Star. https://www.thestar.com/news/gta/2017/04/26/critics-of-new-dynamic-disability-symbol-not-just-anti-pc-cranks-teitel.html

Thomson, R. G. (1997). *Extraordinary bodies: Figuring physical disability in American culture and literature.* Columbia University Press.

Voulgaris, C. T., Smart, M. J. & Taylor, B. D. (2017). Tired of Commuting? Relationships among Journeys to School, Sleep, and Exercise among American Teenagers. *Journal of Planning Education and Research,* 39(2), 142–154. https://doi.org/10.1177/0739456×17725148

Wasserman, D. (1998). Distributive Justice. In A. Silvers, D. Wasserman, & M. B. Mahowald (Eds.), Disability, Difference, and Discrimination: Perspectives on Justice in Bioethics and Public Policy (p. 147-207)

Wittgenstein, L. (2009). *Philosophical investigations.* (G. E. M. Anscombe, P. M. S. Hacker, and J. Schulte, Trans.) Wiley-Blackwell. (Original work published 1953)

Author Biographies

Maha Bali – *American University in Cairo*

Maha Bali is Professor of Practice at the Center for Learning and Teaching at the American University in Cairo. She has a PhD in Education from the University of Sheffield, UK. She is co-founder of virtuallyconnecting.org (a grassroots movement that challenges academic gatekeeping at conferences) and co-facilitator of Equity Unbound (an equity-focused, open, connected intercultural learning curriculum, which has also branched into academic community activities Continuity with Care and Socially Just Academia, and a collaboration with OneHE: Community-building Resources. Most recently, she co-organized the Mid-Year Festival (MYFest) via Equity Unbound, a way of re-imagining professional learning online as nourishing, equitable, emergent, communal, creative and agentic. She writes and speaks frequently about social justice, critical pedagogy, and open and online education. She blogs regularly at http://blog.mahabali.me and tweets @bali_maha

Kya Bezanson

I have fetal alcohol syndrome disorder. I originally came from a family of 10 kids. I am on the Board of Directors for Inclusion BC, an organization that advocates for people with intellectual and developmental disabilities. I am also a co-chair of Inclusion BC's Self-Advocate Committee. I graduated from Kwantlen Polytechnic University (KPU) with my Certificate of Arts. I love to draw and hope to be able to come back to KPU and start my degree in visual arts.

Martha Fay Burtis – *Plymouth State University*

Martha Fay Burtis, is the associate director and learning developer at the Open Learning and Teaching Collaborative at Plymouth State University. In this role, she supports faculty with instructional design and pedagogical innovation. Prior to coming to PSU, she was the founding director of the Digital Knowledge Center at the University of Mary Washington. At UMW, she also administered faculty and student development projects, including the Online Learning Initiative and Domain of One's Own. She has particular expertise and interest in digital literacy and pedagogy; student-centered teaching and learning; and critical instructional design.

Catherine J. Denial - *Knox College*

Cate Denial is the Bright Distinguished Professor of American History, Chair of the History department, and Director of the Bright Institute at Knox College in Galesburg, Illinois. A Distinguished Lecturer for the Organization of American Historians, Cate is the winner of the American Historical Association's 2018 Eugene Asher Distinguished Teaching award, and a former member of the Digital Public Library of America's Educational Advisory Board. Cate currently sits on the boards of the Western Historical Quarterly and Commonplace: A Journal of Early American Life. Cate is at work on a new book, *A Pedagogy of Kindness*, under contract with West Virginia University Press. Her historical research has examined the early nineteenth-century experience of pregnancy, childbirth and child-rearing in Upper Midwestern Ojibwe and missionary cultures, research that grew from Cate's previous book, *Making Marriage: Husbands, Wives, and the American State in Dakota and Ojibwe Country* (2013). In summer 2018, Cate was an Andrew W. Mellon Fellow at the American Philosophical Society in Philadelphia, PA.

Rossel-Joyce Garcia – *University of California, San Diego*

Rossel-Joyce Garcia is a class of 2021 graduate from UC San Diego, where she received undergraduate degrees in Ethnic Studies and History.

Logan Gorkov – *University of California, San Diego*

Logan Gorkov is a recent graduate from UC San Diego with a BA in History & Studio Art. He is particularly interested in indigenous histories, decolonization, and the end of capitalism. He has a love-hate relationship with academia but believes everyone should have access to education. Outside of the formal academic sphere (for now), he is working a corporate 9-5 with Ajna.com writing about bias, hiring practices, and the state of the world at work (and slightly irresponsibly treats their social media like his own).

Jennifer Hardwick

Jen Hardwick is a dyslexic educator, a settler scholar, and a collaborator for both Including All Citizens and Transforming Accessibility Services projects. She is Chair of the Policy Studies program and a faculty member (cross-appointed) in the Department of English at Kwantlen Polytechnic University on unceded Coast Salish Territories. Jen's interdisciplinary research focuses on settler colonial policy, inclusive education in policy and practice, and Indigenous literary, media, and performance arts. She tweets sporadically at @Jen_Hardwick

Surita Jhangiani – *University of British Columbia*

Surita Jhangiani is an assistant professor in the Faculty of Education at the University of British Columbia, Point Grey Campus, which is situated on the traditional, ancestral, unceded territory of the Musqueam people. She primarily teaches in the Bachelor of Education program and also teaches upper level graduate courses in the area of Human Development, Learning and Culture. Her research interests include scholarship related to open education, alternative grading, pedagogy of care and mental health. Her work is informed by a post-colonial, diversity and gender lens.

Andrew David King – *UC Berkeley*

Andrew (andy) David King is a doctoral student in English at UC Berkeley, where they focus on poetry, philosophy, disability studies, and the history and theory of creative writing pedagogy. They hold an MA in Philosophy from Central European University, where they completed a thesis on accessibility, aesthetics, and public arts funding and served as the student representative on the Committee on Students with Disabilities. They also hold an MFA from the Iowa Writers' Workshop, where they were a Teaching-Writing Fellow, and served in 2019 and 2020 as Provost's Visiting Writer and Visiting Assistant Professor at the University of Iowa Department of English and Research Assistant at the Walt Whitman Archives. Their critical and creative work has appeared or will appear in The Routledge Handbook of Ecofeminism and Literature (Routledge, 2022), A Field Guide to the Poetry of Theodore Roethke (Swallow Press/Ohio University Press, 2020), The Scholarship of Creative Writing Practice: Beyond Craft, Pedagogy, and the Academy (Bloomsbury, forthcoming), Best New Poets 2018 and 2020 (University of Virginia Press), and a forthcoming volume on pedagogy and the book arts.

Mary Klann – *University of California, San Diego*

Mary Klann is an adjunct lecturer in San Diego, where she teaches courses in Native American History, Digital History, Women's History, and US History at UC San Diego and San Diego Miramar College.

Her first book, Wardship and the Welfare State: Native Americans and the Formation of First-Class Citizenship in Mid-Twentieth-Century America, is under contract with University of Nebraska Press. Her writing has also appeared in the Journal of Women's History, Women and Social Movements, Smithsonian Magazine, Contingent Magazine, and Crowdsourcing Ungrading. In addition to teaching and research, she also enjoys working with faculty to facilitate conversations about inclusive pedagogy and online teaching. She tweets occasionally at @mcklann, mostly about issues relating to Native history, ungrading, and adjuncting.

Pat Lockley

hopes you enjoy reading his chapter and is sorry if you don't.

Anju Miller

My biological name is Katie, and online I'm known as Sarafine but my preferred name is Anju. I'm a graduate of Kwantlen Polytechnic University and the Including All Citizens and am currently pursuing a creative writing bachelor's degree. I love writing and creating characters and new laws and landscapes for them to live in. Currently, I work at Toys R Us and have been with the company for eight years. All my life I have been told by several critical people in my life "You can't do____because you have a disability" but I am determined to help educate the people who think like that and show them not to put limitations on a person because of their disability. My goal is to increase an understanding around the disability community across Canada first then eventually the world. Together we are strong but being segregated brings weakness.

Jessica O'Reilly

is human, like you.

Nicolas Armando Parés

Nicolas Parés is an instructional designer and teaches courses in the Colorado community college system. He serves on the State of Colorado's Open Educational Resources (OER) Council and focuses his re search on reducing barriers to education for culturally and linguistically diverse learners and adult learners.

Jerod Quinn – *Wake Forest University*

Jerod Quinn is an instructional designer at Wake Forest University in the Office of Online Education. He's been an ID for well over a decade helping instructors create online classes they are excited to teach and that learners are excited to take. He has a Master's Degree in Learning Systems Design and Development and is a dissertation away from a PhD in Educational Psychology with an emphasis in Quantitative Measurement from the University of Missouri. His work bubbles out of the tension of holding the pragmatic (what can actually be done) in one hand while holding the ideal (what should be done) in the other. His first book, The Learner Centered Instructional Designer, came out in 2021 through Stylus Publishing. You can find him tweeting about teaching, learning, instructional design, guitars, and nature at @jerodq.

Benjamin D. Remillard – *University of New Hampshire*

Ben is a PhD Candidate at the University of New Hampshire. In addition to writing about digital pedagogy, his research focuses on the lives of the American Revolution's veterans of color and their experiences in the early republic. He lives in Massachusetts with his wife, daughter, and hedgehog.

Emma Sawatzky

I was diagnosed with ADHD when I was eight years old. After I graduated from high-school, I enrolled in Kwantlen Polytechnic University's (KPU) Adult Special Education employment-based program, Access Program for People with Disabilities. Because of this program I met Fiona and I joined her film club, The Bodies of Film Club. We watch movies that have characters with disabilities, and we wrote a chapter for the Routledge Companion to Disability and Media. I am now a graduate of KPU and Including All Citizens. It is here where I met my two closest friends. My goal is to go back to school for Early Childhood Education so that I can be a preschool teacher.

Mandi Singleton – *University of Denver*

Mandi Singleton is a teaching and learning specialist and adjunct faculty at University College. She holds a certificate in Educational Leadership and Policy and was a middle school dean, instructional coach, and math and science teacher for 15 years. She has taught problem-based learning (STEM) education in Shanghai, China, and participated in the Colorado BioScience Institute Research Experience for Teachers on the Anschutz Medical Campus with Sharklet Technologies and Dr. Chelsea Magin. While there, she became a contributor on a science publication and was named Educator of the Year in 2015. You can find her at the Anschutz Health and Wellness Center gym picking things up and putting them down while training for powerlifting or advocating for more teacher development focus in the higher education course delivery space.

Laurel Staab

Laurel Staab is a learning designer and practitioner-based researcher committed to educational justice. She works as a Senior Learning Designer at Multiverse, an education startup that provides professional apprenticeships as an alternative to higher education. She previously worked as the Director of Innovative Learning and Pedagogy at African Leadership University where she was a founding faculty member in Mauritius and led the founding faculty team in Rwanda. She is passionate about finding and creating collaborative spaces where learners have autonomy over their own experiences. She tweets occasionally @ lastaab about education, pic abolition, politics, and bad movies.

Martin Wairimu

Martin Wairimu is an educator passionate about learning design and decolonizing education. He has a BA(hons) in Global Challenges from the African Leadership University (ALU), Rwanda. He is currently working with the United Nations Institute of Training and Research (UNITAR), in the learning solutions team, developing modules on gender and inclusivity for trainers of trainers improving UNITAR's quality of training and projects under the division for peace. Martin has previously worked at the office of the Director of Innovative Learning and Pedagogy at ALU, where he designed curriculum and learning experiences that leverage innovative, hands-on and student centered methodologies. Additionally, they reviewed curriculum and learning objectives for a US Department of Defense-funded four-week program titled 'Securing the State – Building Institutions for National Security," and most recently provided administration and logistical support for the African Union Staff Officer Course (AUSOC) in support of a US-AFRICOM Justified Accord 22 Exercise in Nairobi. Martin is joining UNITAR as a trainee to support the Learning Solutions team within the Division for Peace.

Fiona Whittington-Walsh – *Kwantlen Polytechnic University*

I am an instructor in the sociology department and director of the Including All Citizens at Kwantlen Polytechnic University (KPU) in British Columbia, Canada. I am also KPU's Lead Advisor on Disability, Accessibility, and Inclusion. My current research projects include disability and film, inclusive pedagogy, and transforming post-secondary services for students with disabilities. A key aspect to my work is creating strong connections with the community. I sit on several provincial and national Boards of Directors for non-profit organizations that advocate for the full inclusion of people with intellectual and/or developmental disabilities. These include Inclusion BC, Inclusion Canada, and the Institute for Research and Development on Inclusion and Society (IRIS).

Mia Zamora – *Kean University*

Mia Zamora, Ph. D. is Professor of English, Director of the MA in Writing Studies & the Kean University Writing Project in Union, NJ, USA. A recent recipient of Kean University's "Professor of the Year" Award, Zamora is a scholar of Electronic Literature. Dr. Zamora's commitment to equity, digital literacies, care, and intercultural understanding is clear in both her public scholarship and leadership work. As a leading voice for the practice of open networked education, she has founded several global learning networks including Equity Unbound (#unboundeq) and Networked Narratives (#netnarr). She currently is co-organizing #MYFest22 – a Mid-Year Festival "recharge and renewal experience" exploring open educational practices, open publishing, socially just education, community building reflection, and wellbeing/joy

www.ingramcontent.com/pod-product-compliance
Lightning Source LLC
Chambersburg PA
CBHW050637160426
43194CB00010B/1707